DATA ANALYTICS IN BIOMEDICAL ENGINEERING AND HEALTHCARE

DATA ANALYTICS IN BIOMEDICAL ENGINEERING AND HEALTHCARE

Edited by

KUN CHANG LEE
Professor, SKK Business School, Sungkyunkwan University, Seoul, Korea

SANJIBAN SEKHAR ROY
Associate Professor, School of Computer Science and Engineering, VIT University, Vellore, Tamil Nadu, India

PIJUSH SAMUI
Associate Professor, Department of Civil Engineering, NIT Patna, Patna, Bihar, India

VIJAY KUMAR
Associate Professor, School of Electronics Engineering, VIT University, Vellore, Tamil Nadu, India

Academic Press is an imprint of Elsevier
125 London Wall, London EC2Y 5AS, United Kingdom
525 B Street, Suite 1650, San Diego, CA 92101, United States
50 Hampshire Street, 5th Floor, Cambridge, MA 02139, United States
The Boulevard, Langford Lane, Kidlington, Oxford OX5 1GB, United Kingdom

© 2021 Elsevier Inc. All rights reserved.

No part of this publication may be reproduced or transmitted in any form or by any means, electronic or mechanical, including photocopying, recording, or any information storage and retrieval system, without permission in writing from the publisher. Details on how to seek permission, further information about the Publisher's permissions policies and our arrangements with organizations such as the Copyright Clearance Center and the Copyright Licensing Agency, can be found at our website: www.elsevier.com/permissions.

This book and the individual contributions contained in it are protected under copyright by the Publisher (other than as may be noted herein).

Notices
Knowledge and best practice in this field are constantly changing. As new research and experience broaden our understanding, changes in research methods, professional practices, or medical treatment may become necessary.

Practitioners and researchers must always rely on their own experience and knowledge in evaluating and using any information, methods, compounds, or experiments described herein. In using such information or methods they should be mindful of their own safety and the safety of others, including parties for whom they have a professional responsibility.

To the fullest extent of the law, neither the Publisher nor the authors, contributors, or editors, assume any liability for any injury and/or damage to persons or property as a matter of products liability, negligence or otherwise, or from any use or operation of any methods, products, instructions, or ideas contained in the material herein.

Library of Congress Cataloging-in-Publication Data
A catalog record for this book is available from the Library of Congress

British Library Cataloguing-in-Publication Data
A catalogue record for this book is available from the British Library

ISBN 978-0-12-819314-3

For information on all Academic Press publications
visit our website at https://www.elsevier.com/books-and-journals

Publisher: Conner, Mara
Acquisitions Editor: Katsaropoulos, Chris
Editorial Project Manager: Silva, Isabella C.
Production Project Manager: Raviraj, Selvaraj
Cover Designer: Bilbow, Christian J.

Typeset by SPi Global, India

I would like to dedicate this book to my son Sanjith
—Sanjiban Sekhar Roy

Contents

Contributors xiii
Preface xvii

Chapter 1 Estimate sequential poses for wireless endoscopic capsule based on encoder-decoder convolutional neural network toward autonomous navigation
Navid Panchi, Mengya Xu, Mobarakol Islam, Archana Gahiwad, and Hongliang Ren

1 Introduction 1
2 System overview and analysis 3
3 Dataset 6
4 Training and evaluation 6
5 Evaluation and results 7
6 Conclusions 9
Acknowledgments 10
References 10

Chapter 2 Interoperability issues in EHR systems: Research directions
Sreenivasan M. and Anu Mary Chacko

1 Introduction 13
2 Types of interoperability 15
3 Interoperability in healthcare 18
4 Use case of semantic interoperability in healthcare 19
5 Benefits of semantic interoperability 21
6 Challenges in semantic interoperability 22

7 Research directions 23
8 Conclusion 23
References 27

Chapter 3 Difficulty with language comprehension and arithmetic word problems due to hearing impairment: Analysis and a possible remedy through a new Android-based assistive technology
Sandipa Roy, Arpan Kumar Maiti, Gopal Krishna Basak, and Kuntal Ghosh

1 Introduction 29
A Statistical analysis toward understanding the problem 31
B Possible solution through app-based technology 31
A Statistical analysis toward understanding the problem 32
2 Description of the problem 32
3 Survey methodology 35
4 Statistical analysis of the survey outcome 40
B Possible solution through app-based technology 43
5 App requirement analysis 43
6 App design 45
7 Results and discussion 48
8 Conclusion 52
Acknowledgments 54
References 55

Chapter 4 Machine learning in healthcare toward early risk prediction: A case study of liver transplantation
Parag Chatterjee, Ofelia Noceti, Josemaría Menéndez, Solange Gerona, Melina Toribio, Leandro J. Cymberknop, and Ricardo L. Armentano

1 Introduction 57
2 Background of the study: Description of cohort 60
3 Intelligent risk analysis: Aspect of vascular age and cardiometabolic health 62

4 Results and discussions 67
5 Conclusion 70
Acknowledgments 71
References 71

Chapter 5 Utilizing BERT for biomedical and clinical text mining
Runjie Zhu, Xinhui Tu, and Jimmy Xiangji Huang

1 Introduction and motivation 73
2 BERT 75
3 Biomedical and clinical BERT 76
4 Conclusions and future work 98
Acknowledgment 99
References 99

Chapter 6 Classifying CT scan images based on contrast material and age of a person: ConvNets approach
Soumik Mitra, Sanjiban Sekhar Roy, and Kathiravan Srinivasan

1 Introduction 105
2 Objectives 106
3 Convolutional neural networks 106
4 Related study 108
5 Dataset 109
6 Methodology 110
7 Results and discussion 114
8 Conclusion 116
References 116

Chapter 7 Data analytics in IOT-based health care
Azadeh Zamanifar

1 Introduction 119
2 Off-line data analytic in IoT-based health care 121
3 Online data analytic in IoT-based health-care systems 124

 4 Open research and challenges 126
 5 Conclusion 126
 References 127

Chapter 8 Application of PCA based unsupervised FE to neurodegenerative diseases
 Y.-H. Taguchi and Hsiuying Wang

 1 Introduction 131
 2 Materials and methods 132
 3 Results 134
 4 Discussions 143
 References 143

Chapter 9 Disease diagnosis using machine learning: A comparative study
 Rakshit Jain, Asmita Chotani, and G Anuradha

 1 Introduction 145
 2 Related work 147
 3 Methodology 149
 4 Dataset and metrics 155
 5 Result 156
 6 Conclusion 158
 References 159

Chapter 10 Driver drowsiness detection using heart rate and behavior methods: A study
 Anmol Wadhwa and Sanjiban Sekhar Roy

 1 Introduction 163
 2 Heartbeat detection method 165
 3 Detecting driver drowsiness using behavioral method 168
 4 Conclusion 173
 References 173

Chapter 11 Innovative classification, regression model for predicting various diseases
B.K. Tripathy, M Parimala, and G Thippa Reddy

1 Introduction 179
2 Classification 180
3 Regression model for disease prediction 189
4 Challenges in predicting disease 198
5 Conclusion 199
References 199
Further reading 203

Chapter 12 Clavicle bone segmentation from CT images using U-Net-based deep learning algorithm
Parita Sanghani, Francis Wong, and Hongliang Ren

1 Introduction 205
2 Materials and methods 206
3 Results and discussion 210
4 Conclusion 213
Acknowledgements 214
References 214

Chapter 13 Accurate classification of heart sounds for disease diagnosis by using spectral analysis and deep learning methods
Pratima Upretee and Mehmet Emin Yüksel

1 Introduction 215
2 Method 218
3 Performance evaluation 224
4 Results and discussion 225
5 Conclusion 229
References 229

Chapter 14 Complex neutrosophic δ-equal concepts and their applications in water quality
Prem Kumar Singh

1 Introduction 233
2 Three-way complex neutrosophic set and its graph 238
3 Proposed section 244
4 Illustration 251
5 Discussions 260
6 Conclusions 264
Acknowledgments 264
References 264

Index 269

Contributors

G Anuradha Vellore Institute of Technology, Vellore, India

Ricardo L. Armentano National Technological University (Universidad Tecnológica Nacional), Buenos Aires, Argentina; University of the Republic (Universidad de la República), Montevideo, Uruguay

Gopal Krishna Basak Indian Statistical Institute, Kolkata, India

Anu Mary Chacko National Institute of Technology Calicut, Kattangal, Kerala, India

Parag Chatterjee National Technological University (Universidad Tecnológica Nacional), Buenos Aires, Argentina; University of the Republic (Universidad de la República), Montevideo, Uruguay

Asmita Chotani Vellore Institute of Technology, Vellore, India

Leandro J. Cymberknop National Technological University (Universidad Tecnológica Nacional), Buenos Aires, Argentina

Archana Gahiwad NUSRI, Suzhou, China; Department of Biomedical Engineering, National University of Singapore, Singapore, Singapore; VNIT, Nagpur, India

Solange Gerona Military Hospital (Dirección Nacional de Sanidad de las Fuerzas Armadas), Montevideo, Uruguay

Kuntal Ghosh Indian Statistical Institute, Kolkata, India

Jimmy Xiangji Huang School of Information Technology, York University, Toronto, ON, Canada

Mobarakol Islam VNIT, Nagpur, India

Rakshit Jain Vellore Institute of Technology, Vellore, India

Sreenivasan M. National Institute of Technology Calicut, Kattangal, Kerala, India

Arpan Kumar Maiti Indian Statistical Institute, Kolkata, India

Josemaría Menéndez Military Hospital (Dirección Nacional de Sanidad de las Fuerzas Armadas), Montevideo, Uruguay

Soumik Mitra Vellore Institute of Technology, Vellore, India

Ofelia Noceti Military Hospital (Dirección Nacional de Sanidad de las Fuerzas Armadas), Montevideo, Uruguay

Navid Panchi NUSRI, Suzhou, China; Department of Biomedical Engineering, National University of Singapore, Singapore, Singapore; VNIT, Nagpur, India

Parag Chatterjee National Technological University (Universidad Tecnológica Nacional), Buenos Aires, Argentina; University of the Republic (Universidad de la República), Montevideo, Uruguay

M Parimala School of Information Technology and Engineering, VIT Vellore, Vellore, Tamil Nadu, India

G Thippa Reddy School of Information Technology and Engineering, VIT Vellore, Vellore, Tamil Nadu, India

Hongliang Ren Department of Biomedical Engineering, National University of Singapore, Singapore, Singapore; NUSRI, Suzhou, China; VNIT, Nagpur, India

Sandipa Roy Indian Statistical Institute, Kolkata, India

Sanjiban Sekhar Roy School of Computing Science and Engineering, Vellore Institute of Technology, Vellore, Tamil Nadu, India

Parita Sanghani Department of Biomedical Engineering, National University of Singapore, Singapore, Singapore

Prem Kumar Singh Amity School of Engineering and Technology, Amity University, Noida, Uttar Pradesh, India

Kathiravan Srinivasan Vellore Institute of Technology, Vellore, India

Y.-H. Taguchi Department of Physics, Chuo University, Tokyo, Japan

Melina Toribio University of the Republic (Universidad de la República), Montevideo, Uruguay

B.K. Tripathy School of Information Technology and Engineering, VIT Vellore, Vellore, Tamil Nadu, India

Xinhui Tu School of Computer Science, Central China Normal University, Wuhan, China

Pratima Upretee Department of Biomedical Engineering, Graduate School of Natural Sciences, Erciyes University, Kayseri, Turkey

Anmol Wadhwa School of Computing Science and Engineering, Vellore Institute of Technology, Vellore, Tamil Nadu, India

Hsiuying Wang Institute of Statistics, National Chiao Tung University, Hsinchu, Taiwan

Francis Wong Department of Orthopedic Surgery, Sengkang General Hospital Singapore, Singapore, Singapore

Mengya Xu NUSRI, Suzhou, China; VNIT, Nagpur, India

Mehmet Emin Yüksel Department of Biomedical Engineering, Faculty of Engineering, Erciyes University, Kayseri, Turkey

Azadeh Zamanifar Computer Engineering Department, University of Science and Culture, Tehran, Iran

Runjie Zhu Information Retrieval and Knowledge Management Research Lab, Lassonde School of Engineering, York University, Toronto, ON, Canada

Preface

The term data analytics can be associated with predictive analysis, business intelligence, and healthcare techniques. But it is the area of medical science only, where the data analytics term has found its heavy use. The qualitative and quantitative nature of advanced data science has impacted positively on the overall outcome. The state-of-the-art techniques of data analytics can immensely improve the health of the patients. The older techniques of curing patients are getting replaced by the new generation medical equipment that are powered with AI, machine learning, and deep learning technologies. The healthcare institutes are now focusing on value-based payment methods by analyzing the available historical data of the patients and by improving the performance of medical expert systems. The profound use of data analytics has raised the treatment level to a new height. Alongside, it has also reduced the cost to a certain level that boosted the overall profit of the healthcare institutions and efficacy.

Data analytics in healthcare refers to the aggregation of mass data collected from various sources and analyzing them to find out significant insights and information contained therein. The whole process is backed up by up-to-date technology and software powered with various advanced AI techniques. Nowadays the healthcare analytics gradually is understood in a better way than before. This surely ensures positive change of the overall experience of the patient and the quality of care that they face now.

This edited book has dealt with following chains of works on healthcare analytics:
- wireless endoscopic capsule based on autoencoder for autonomous navigation,
- dealing with interoperability issues in EHR,
- analysis and a possible remedy through a new android-based assistive technology for hearing impairment,
- computational intelligence in liver transplantation,
- utilizing BERT for biomedical and clinical text mining,
- deep learning approach for classifying CT scan images,
- data analytics in IOT-based healthcare,
- PCA-based application to neurodegenerative diseases,
- disease diagnosis using machine learning approaches,

- case study of driver drowsiness detection using heart rate and behavior methods,
- clavicle bone segmentation from CT image deep learning model,
- heart sound classification using deep learning methods,
- application of complex neutrosophic δ-equal for water healthcare.

The purpose of compiling this book is to present a vivid idea about both theory and practice related to the aforementioned applications before the readers by showcasing the usages of data analytics, machine learning, and deep learning in health sciences and biomedical data. Besides the book shall be useful to the people working with big data analytics in biomedical research. It will immensely help people working in medical industries, research scholars in the educational institute, and scientists in medical research labs

We hope that readers will be benefited significantly in learning about the state of the art of data analytics applications in healthcare domain including machine learning, data science, deep learning, and AI.

Keep reading, learning, and inquiring.

Dr. Sanjiban Sekhar Roy
Associate Professor
School of Computer Science and Engineering
Vellore Institute of Technology
Vellore, TN, India
Sept 20, 2020
Email: sanjibansroy@ieee.org

Estimate sequential poses for wireless endoscopic capsule based on encoder-decoder convolutional neural network toward autonomous navigation

Navid Panchi[a,b,c], Mengya Xu[a,c], Mobarakol Islam[c], Archana Gahiwad[a,b,c], and Hongliang Ren[a,c]

[a]NUSRI, Suzhou, China. [b]Department of Biomedical Engineering, National University of Singapore, Singapore, Singapore. [c]VNIT, Nagpur, India

1 Introduction

In the last decades, the research on endoscopy has been growing. The conventional method of endoscopy is very time consuming, painful, and it demands expensive equipment. The wireless capsule endoscopy [1] is a painless and effective technique for the diagnosis of the diseases in the digestive tract. As the endoscopic capsules are pill sized, their movement inside the body is much like a medicinal capsule and less invasive than conventional endoscopy techniques [2]. The disadvantage of the current wireless capsule is the missing information on the location of the capsule relative to the anatomy globally. The current wireless capsule endoscopes can only make passive motion controlled by the peristaltic of the inner organs. There are immense benefits in knowing the motion and locations of the capsules for drug delivery and other medical purposes. Hence, many groups have proposed active, remotely controllable, and addressable capsule endoscope [3]. However, the active motion control needs continuous feedback from a reliable and precise real-time pose estimator. Besides, the further estimation of the capsule's position, orientation, namely pose for short, allows more intuitive navigation and correct diagnosis with patient-specific capsule localization and image mapping onto a global coordinate, as shown in the data samples (Fig. 1).

The research presented in this chapter can be used as a basis for automating this task [4–8]. The basic pose estimation techniques like visual odometry estimate the position of the capsule inside the digestive tract. However, the results in these methods are prone to drifts overtime. In addition to that, for real-time tracking and pose estimation, it demands heavy and costly hardware. These techniques are difficult to implement and get the required results.

Various innovations and researches are going on the localization and real-time mapping for all this purpose. There are modern techniques like magnetic resonance imaging (MRI), ultrasonic imaging, and radioactive imaging, among others, which require costly sensors and hardware. The shared disadvantage of these localization methods is that such extra sensors have their limitations when it comes to space limitations, cost, and biocompatibility issues.

In the last decade, machine learning algorithms have produced state-of-the-art results on computer vision-related tasks, like image recognition, object detection, and other prediction problems. Samui et al. [9] and Balas et al. [10] give a comprehensive view on applications of deep learning in different fields. Feature representations form an essential component of machine learning algorithms. Traditionally, many machine learning algorithms used to take handcrafted features as inputs. Considering machine learning models and computer vision techniques, using the neural network approach gives an excellent accuracy without the need for handcrafted features. The motivation for this work was taken from self-driving cars, considering the capsule as a car, we can think of this problem as predicting the steering angle for controlling the vehicle. We have used the encoder-decoder convolutional neural network (CNN) as our model. The dataset of RGB images of real pig stomach is for training and testing purposes. To predict translational and rotational values for the desired

 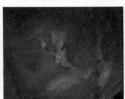 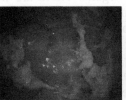 

Fig. 1 Dataset samples.

trajectory of the wireless endoscopic capsule, this proposed model successfully achieved an accuracy of around 99% for the prediction of the 6 DOF values of the capsule. Our model predicts the following next position of the capsule in the digestive tract rather than the current frame position. This feature of the model can make a wireless endoscopic capsule autonomous.

As an outline of this chapter, we first explain the related work and our proposed model. Then, after a brief description of the dataset, the training of the model is explained. The results of our network are presented with the conclusion and future directions.

2 System overview and analysis

2.1 Related work

CNNs have been promising for the detection of different types of diseases or anomalies in the endoscopic images [11, 12] as well as for segmentation of blood vessels in fundus images [13]. Researchers have also used CNNs for the prediction of the 6 DOF pose of the camera using a single RGB image. The neural network architectures proposed in PoseNet [14], PoseCNN [15] trained models on different RGB images containing various objects. In [16], the authors have predicted 6 DOF values of a camera in seven different scenes.

Researchers have been using visual odometry for the prediction of the pose of the endoscopic capsule in the past. The comparison with different visual odometry techniques and a neural network is in Turan et al. [5]. The most recurring problem with visual odometry is the need for a textured surface, which the digestive tract may lack at times. In [17], they have used inception architecture with LSTM layers at the end instead of fully connected layers. Their model is trained on depth images obtained using the TFS-Shah method. Since their architecture is computationally costly to infer in real time, it is not possible to use it for real-time pose prediction. Also, the network requires depth images with the RGB images, which is also a big drawback. In [18], they removed the softmax layer from the GoogLeNet and added a fully connected layer to regress 6 DOF pose values.

In [19], the authors have proposed using two networks for unsupervised posed estimation, the presence of supervised ground-truth labels, there is no need for unsupervised learning. Also, the usage of two networks makes the model to require high computational power, which can make it difficult for real-time inference. The usage of a single neural network in our proposed method also makes it easier if one needs to use model compression to reduce the computational costs during deployment of the network.

The prediction accuracy of the encoder-decoder CNN depends on the representation of the features. Although in deep neural networks, we do not explicitly define features, we need to make sure that the network is learning prominent features, for the network to give accurate results on the given task.

In our work, we propose an architecture similar to the hourglass network [16]. The encoder part of the network learns robust feature representation. If we used a simple feed-forward network, then the "where" information is lost, and while predicting the 6 DOF values, the accuracy was not very good. It loses the "where" information due to the max pooling layer, as the pooling layers are to make the output of the model somewhat invariant to the input rotations and translations. It is different in our case because our output is very much dependent on input rotations and translations. To solve this problem, the authors of [16] proposed using an encoder-decoder neural network with fully connected layers at the end for the regression of the pose values. They have called the network hourglass pose network. We propose the usage of the same network to solve our problem. The comparison is given in the following section. We can train the network directly on RGB images and predict the pose, which the capsule has to acquire next. The proposed system takes the endoscopic image and regresses the camera's 6 DOF pose without a need for any extra sensor.

2.2 Method

The main aim of this work is the estimation of the subsequent pose of the capsule directly from an RGB image. We utilize a CNN architecture for prediction of seven values, in the form of two vectors, a three-dimensional (3D) translation vector, and a four-dimensional (4D) rotation (quaternion) vector. The overall network architecture in Fig. 2 consists of three

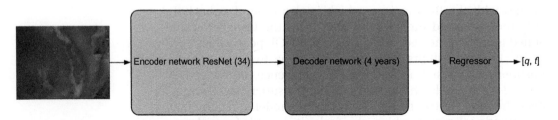

Fig. 2 Overview of architecture.

components: an encoder, a decoder, and a regressor. The encoder acts as a feature extractor, while the decoder recovers the relevant details of the input from the encoder outputs. Eventually, the regressor estimates the capsule pose, taking in the production of the decoder.

2.3 Proposed model

It is a common practice to work out CNNs for different computer vision tasks. A nine-layer CNN was not able to learn properly due to less number of parameters. Then, we use the network proposed by Turan et al. [17]. We then use the inception network by replacing the last layer with a linear layer having seven output units, objective function used for optimizing the parameters was MSE loss [12], this network was unable to generalize over the test set. We intentionally kept the network small to ensure less inference time, while the less number of learnable parameters learning curve does not saturate after a few epochs. The encoder-decoder architecture similar to the architecture proposed in [16] turned out to be best among all the architectures that we investigated. The details of this network are given below.

A modified ResNet34 [21] architecture constitutes the encoder part of the network. The decoder part of the network is using three up-convolutional layers, followed by one convolutional layer. The ResNet34 architecture removes the nonconvolutional part of the architecture, that is, removal of the pooling as well as the fully connected layers from the end. The encoder part reduces the spatial resolution of the feature maps, while the up-convolution layers in the decoder part gradually increase it. The convolutional layer is for dimensionality reduction. The purpose of using the decoder part is the restoration of the essential fine-grained visual information of the input image, which was lost during the convolution operations. The skip connections between the encoder blocks and the corresponding layers in the decoder layers ensure the reutilization of features from the earlier layers of the network [16]. The output from the decoder is flattened and then passed to a fully connected layer with 2048 units. It is passed on to two different affine regression layers with 4 and 3 units for quaternion and translational vector predictions, respectively. The dataset used for training the model is explained in Section 4. The specifications of the layers are given in Fig. 3. We have used direct RGB images to facilitate an end-to-end learning. PyTorch [22] framework was used to implement neural networks.

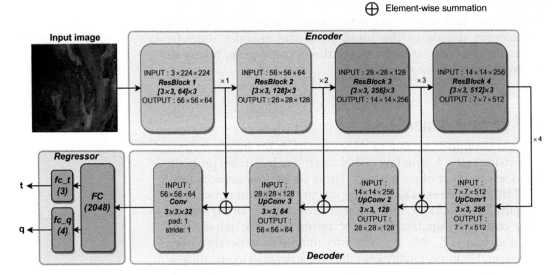

Fig. 3 Hour glass encoder-decoder model [20].

3 Dataset

A dataset containing 12,000 RGB images (640 × 480 resolution) with proper 6 DOF (four quaternion elements and three translation elements) values as labels were used for training, testing, and validation. Out of 12,000 images, 10,000 images were used to train the network, and 1000 images were used for validation purposes and rest for testing purposes. The testing accuracy is discussed in the results section. We have not used any depth images as used in DeepEndoVo [17] to make sure that there is no preprocessing step during the real-time inference, which can create lag.

The dataset is collected with the magnetically actuated soft capsule endoscopes (MASCE). The images were taken from a pig's stomach with various arrangements. The dataset used by us is the same as in DeepEndoVo [17]. For further details on how the dataset was acquired, the reader is directed to Ref. [17].

4 Training and evaluation

4.1 Training

For all of our models except for the nine-layered CNN, we initialized the weights of the layer to the weights pretrained on ImageNet [23] to have a better initialization rather than using a random initialization. All the networks were trained using Adam optimizer with a learning rate of 0.001 for the first 100 epochs and then 0.0001 for the next 100 epochs. The loss function used is L2 Euclidean loss. A scaling factor, β, decides the trade-off between the rotational and translational values. It was initially 500 for the first 100 epochs and 100 for the next 100 epochs. The neural networks were trained for 200 epochs and a few more epochs with hyperparameter tuning. The training algorithm is given in Algorithm 1.

Training algorithm

loop:

- Take one batch of RGB images from the training dataset
- Normalize the image matrices
- Convert the image matrices to tensors
- Take the loss between ground-truth and predicted values
- Compute gradients of the weights
- Back propagate
- **if** *loss* < *loss_threshold* **then** break

The training and validation losses are in Fig. 4. It can also be seen the loss eventually converges, and the validation loss is nearly the same as the training loss, through which we can conclude that there is no overfitting of the network, which would otherwise affect the ability of the network to generalize.

5 Evaluation and results

The loss function is given by

$$loss(I) = \|\hat{x} - x\|_2 + \beta \|\hat{q} - q\|_2 \qquad (1)$$

where (x, q) are the ground-truth position-rotation pairs and $(\hat{x}, \hat{q})$ are estimated position-rotation pairs, respectively. β is a scale factor that can be tuned by grid search.

Before the image is fed into the network, all the images of the evaluation dataset are resized to (224×224) pixels. This is necessary as the network is trained for prediction of pose using (224×224) resolution images. As we could not fit lots of high-resolution images in GPU memory, the above step was essential. Image normalization is also performed before training as well as evaluation. The average mean and average standard deviation of the pixel values in all the images is calculated beforehand for the entire

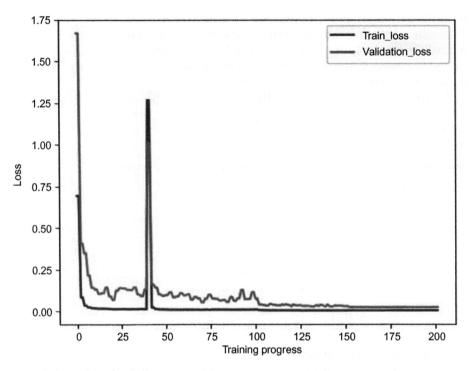

Fig. 4 Loss plot during training (scaled).

training dataset. These mean and standard deviation values are used for the normalization of both training and evaluation datasets.

All models were trained on NVIDIA K5200 GPU, and the training was performed using mini-batches of size 34. A dropout layer followed every fully connected layer with a dropout probability of 0.5. This helped us in reducing the overfitting. Batch normalization was used for avoiding overfitting.

After training, the network was tested using data that the network has never seen during training. The errors and accuracy are given in the results section.

5.1 Testing

Accuracy can be defined as the percentage ratio of average error in a particular value and its standard deviation are as follows:

$$Accuracy(X) = avg \frac{error\,(X)}{std\,deviation\,(X)} \times 100 \qquad (2)$$

The results are presented in different formats. For comparison, we have given ground-truth labels and predicted subsequent pose in Table 1. The accuracy for every value is given in Table 2. The maximum error in translation values and rotational values were found to be 3.5 mm and 0.006 degrees, respectively. The overall accuracy of values is 99.45, and for translation values, it is 99.22%.

6 Conclusions

In this research work, the authors present an approach based on encoder-decoder CNN, which can be trained in the end-to-end manner for image-based autonomous navigation of a wireless endoscopic capsule. The neural network gives 6 DOF values when fed with an RGB image from the camera of the capsule using the hourglass-sum network, without image preprocessing except for resizing and normalization. This can be used in real time by reducing the inference time and increasing the frame rate per second. At last, the results are presented in various forms in terms of accuracy with a minimal error margin. Future work can be done

Table 1 Comparison between ground-truth values and predicted values.

	Rotational	Rotational.1	Rotational.2	Rotational.3	Position	Position.1	Position.2
Ground-truth values	0.1704	−0.1350	−0.2462	−0.9384	−395.543	200.8725	734.6886
Predictions	0.1756	−0.1415	−0.2474	−0.9423	−391.4963	194.4396	729.7900
Ground-truth values	0.1721	−0.0463	−0.2905	−0.9201	−336.1646	224.2240	665.2360
Predictions	0.1726	−0.0484	−0.2821	−0.9425	−341.6953	226.3347	675.3682
Ground-truth values	0.2555	0.1814	−0.1416	−0.9275	−336.5125	197.1444	555.8495
Predictions	0.2756	0.1934	−0.1338	−0.9321	−333.7270	198.7227	548.4428
Ground-truth values	0.2455	−0.2006	−0.3218	−0.8752	−340.8595	204.3742	593.6349
Predictions	0.2553	−0.2007	−0.3124	−0.8927	−345.2036	206.0328	597.3804
Ground-truth values	0.2037	−0.0358	−0.1746	−0.9543	−393.8181	146.6441	689.3059
Predictions	0.2120	−0.0386	−0.1651	−0.9625	−389.7674	143.0319	689.5455

Table 2 Average error for rotational and translational values.

	Rotational	Rotational.1	Rotational.2	Rotational.3	Position	Position.1	Position.2
Error (degrees/mm)	0.000605	0.005697	0.002738	0.005094	2.151201	0.931700	3.532875
Accuracy (%)	99.87	99.55	99.45	98.94	99.36	99.53	98.76

on decreasing the loss and using model compression techniques for improving the frame rate during inference.

Acknowledgments

This work is partially supported by the National Key Research and Development Program, the Ministry of Science and Technology (MOST) of China (No. 2018YFB1307703). The authors would like to acknowledge the authors of DeepEndoVO [17] for providing us with their dataset for this research work.

References

[1] G. Iddan, G. Meron, A. Glukhovsky, P. Swain, Wireless capsule endoscopy. Nature 405 (6785) (2000) 417, https://doi.org/10.1038/35013140.
[2] Z. Liao, R. Gao, C. Xu, Z.-S. Li, Indications and detection, completion, and retention rates of small-bowel capsule endoscopy: a systematic review, Gastrointest. Endosc. 71 (2) (2010) 280–286.
[3] T. Nakamura, A. Terano, Capsule endoscopy: past, present, and future, J. Gastroenterol. 43 (2) (2008) 93–99.
[4] M. Turan, Y. Almalioglu, H. Araujo, E. Konukoglu, M. Sitti, A non-rigid map fusion-based direct SLAM method for endoscopic capsule robots, Int. J. Intell. Robot. Appl. 1 (4) (2017) 399–409.
[5] M. Turan, J. Shabbir, H. Araujo, E. Konukoglu, M. Sitti, A deep learning based fusion of RGB camera information and magnetic localization information for endoscopic capsule robots, Int. J. Intell. Robot. Appl. 1 (4) (2017) 442–450.
[6] M. Turan, A. Abdullah, R. Jamiruddin, H. Araujo, E. Konukoglu, M. Sitti, Six degree-of-freedom localization of endoscopic capsule robots using recurrent neural networks embedded into a convolutional neural network, arXiv preprint arXiv:1705.06196 (2017).
[7] M. Turan, Y.Y. Pilavci, R. Jamiruddin, H. Araujo, E. Konukoglu, M. Sitti, A fully dense and globally consistent 3d map reconstruction approach for GI tract to enhance therapeutic relevance of the endoscopic capsule robot, arXiv preprint arXiv:1705.06524 (2017).
[8] M. Turan, Y.Y. Pilavci, I. Ganiyusufoglu, H. Araujo, E. Konukoglu, M. Sitti, Sparse-then-dense alignment-based 3D map reconstruction method for endoscopic capsule robots, Mach. Vis. Appl. 29 (2) (2018) 345–359.
[9] P. Samui, S.S. Roy, V.E. Balas, Handbook of Neural Computation, Academic Press, 2017.

[10] V.E. Balas, S.S. Roy, D. Sharma, P. Samui, Handbook of Deep Learning Applications, vol. 136, Springer, 2019.

[11] H. Alaskar, A. Hussain, N. Al-Aseem, P. Liatsis, D. Al-Jumeily, Application of convolutional neural networks for automated ulcer detection in wireless capsule endoscopy images. Sensors 19 (6) (2019), https://doi.org/10.3390/s19061265.

[12] Y. Yuan, M.Q.-H. Meng, Deep learning for polyp recognition in wireless capsule endoscopy images. Med. Phys. 44 (4) (2017) 1379–1389, https://doi.org/10.1002/mp.12147.

[13] R. Biswas, A. Vasan, S. Roy, Dilated deep neural network for segmentation of retinal blood vessels in fundus images https://rdcu.be/bG8rM. Iranian J. Sci. Technol. Trans. Electr. Eng. (2019), https://doi.org/10.1007/s40998-019-00213-7.

[14] A. Kendall, M. Grimes, R. Cipolla, Posenet: a convolutional network for realtime 6-DOF camera relocalization, in: Proceedings of the IEEE International Conference on Computer Vision, 2015, pp. 2938–2946.

[15] Y. Xiang, T. Schmidt, V. Narayanan, D. Fox, Posecnn: a convolutional neural network for 6D object pose estimation in cluttered scenes, arXiv preprint arXiv:1711.00199 (2017).

[16] I. Melekhov, J. Ylioinas, J. Kannala, E. Rahtu, Image-based localization using hourglass networks, in: 2017 IEEE International Conference on Computer Vision Workshops (ICCVW), 2017, pp. 870–877.

[17] M. Turan, Y. Almalioglu, H. Araujo, E. Konukoglu, M. Sitti, Deep endovo: a recurrent convolutional neural network (RCNN) based visual odometry approach for endoscopic capsule robots, Neurocomputing 275 (2018) 1861–1870.

[18] M. Turan, Y. Almalioglu, E. Konukoglu, M. Sitti, A deep learning based 6 degree-of-freedom localization method for endoscopic capsule robots, (2017.)https://arxiv.org/abs/1705.05435.

[19] M. Turan, E.P. Ornek, N. Ibrahimli, C. Giracoglu, Y. Almalioglu, M.F. Yanik, M. Sitti, Unsupervised odometry and depth learning for endoscopic capsule robots, IEEE/RSJ International Conference on Intelligent Robots and Systems (IROS) (2018), http://arxiv.org/abs/1803.01047.

[20] V. Badrinarayanan, A. Kendall, R. Cipolla, Segnet: a deep convolutional encoder-decoder architecture for image segmentation, arXiv preprint arXiv:1511.00561 (2015).

[21] K. He, X. Zhang, S. Ren, J. Sun, Deep residual learning for image recognition, in: Proceedings of the IEEE Conference on Computer Vision and Pattern Recognition, 2016, pp. 770–778.

[22] S. Ju, S. Ramjee, D. Yang, A. El Gamal, A PyTorch framework for automatic modulation classification using deep neural networks (2018). Thesis.

[23] O. Russakovsky, J. Deng, H. Su, J. Krause, S. Satheesh, S. Ma, Z. Huang, A. Karpathy, A. Khosla, M. Bernstein, Imagenet large scale visual recognition challenge, Int. J. Comput. Vis. 115 (3) (2015) 211–252.

Interoperability issues in EHR systems: Research directions

Sreenivasan M. and Anu Mary Chacko
National Institute of Technology Calicut, Kattangal, Kerala, India

1 Introduction

The primary goal of all healthcare systems is to provide the best healthcare experience to their clients, that is, patients. To provide high-quality, trustworthy healthcare, countries across the globe are working on converting the paper-based documentation to computer-based recordings (electronic health records (EHRs)). Electronic health records are the backbone of health information systems (HISs) and have revolutionized healthcare delivery. Various sources like hospitals, pharmacy, mobile devices, IoT, and social media all generate data that are useful for EHR construction.

Currently, EHRs are generated via data entry done by clinicians, doctors, nurses, etc., treating the patient, and hospitals, where the patient has taken the treatment, have full access to EHR. Patients do not have direct access to their EHR but will be provided part of the information in the form of lab reports and discharge summary. This restriction allows a hospital monopoly and is not in the best interest of the patient. Thus there is a need to create a framework that is patient centric where EHR of a particular patient and all health-related data pertaining to the patient, personal health record (PHR), is available to him and to those whom he wants to give access to. PHR can be enriched by including information on patient reports, laboratory results, and data from smartphones and other health monitoring wearable devices. Advantage of a PHR over traditional EHR is that it ensures storing of patient's medical records in a single place instead of storing them at various hospitals. A complete and accurate summary of an individual's medical history across hospitals and different ailments can thus be generated from PHR. The availability of such consolidated data will help clinicians to provide better treatment by making informed decisions. To generate PHR the fundamental requirement is that EHRs need to be consolidated. There need to be mechanisms where the EHR generated by different sources can be understood and are interoperable.

In the healthcare domain, at any specified stage in time, one user operates in distinct role capacity. For example, a doctor may be a primary physician for one patient and serve as a specialist for another patient. The information in the EHR is confidential and needs to be made available after proper authentication as per authorization. The full power of EHR can only be leveraged in an interoperable environment. Communication, collaboration, and shared care should be established among health information systems to provide better healthcare.

Experts from different domains (like healthcare, information systems and security, computer science, public health, and public policy) have developed various structural and semantic standards to represent, share, and provide a consistent semantic representation for EHR. Interoperability helps in delivering communication and collaboration among healthcare systems and helps in improving the security, safety, and quality of healthcare.

In all distributed systems, interoperability is an essential characteristic and is the ability of computer systems or software to exchange and use information. ISO3 defines interoperability of electronic health record as "the ability of two or more applications being able to communicate effectively without compromising the content of transmitted EHR" [1]. EHR information needs to be shared among hospitals (within or outside sharing), clinicians, laboratories, insurance, etc., without affecting the privacy of the patient. This is challenging as currently it is found that different units/departments work according to different proprietary/open standards that are available like HL7, Fast Healthcare Interoperability Resources (FHIR), openEHR, and Continuity of Care Record (CCR). This results in various issues related to interoperability like naming disputes and resolving dependencies among access control. Syntactic and semantic validation is required to assess such characteristics. It is therefore essential to provide dynamic feedback to the scheme that will continue to run a specific algorithm needed to resolve any potential disputes. Interoperable issues of naming conflicts and resolving dependencies between distinct characteristics of distinct access control strategies also require an assessment of such traits to achieve two separate layers of syntactic and semantic validation.

Interoperability not only is applicable for EHR but also is used in many other areas. Web technology consists of a large variety of complex web-based information services. As web technology spreads, web-based information services are exploding in number. The web-based application allows a user to access web resources through a semantic layer with the function of incorporating multiple data resources in the same or similar domains. This is due to the distributed, heterogeneous, and open nature of these applications. Therefore a web-based application system requires a generic architecture by supporting semantic interoperability to incorporate web resource propagation.

A smart grid is an electricity supply network that monitors and responds to local changes in the use of electricity using digital communication technologies. Smart grid vision requires both syntax interoperability to allow exchange information physically and semantic interoperability to understand and interpret the meaning of the changes. Semantic technology in a smart grid system can be used to make decisions. The issue of a unified global ontology in a smart grid scenario is yet to be achieved [2].

In the healthcare scenario, there is a higher requirement for semantic interoperability as the system is managed by nontechnologists, and most notes are written in natural language where terms are from the medical domain. The existing EHR solutions are complying to different standards, and it is challenging to consolidate and derive meaning from these data. To do an intelligent summarization for semantic interoperability, the use of artificial intelligence is needed and is very challenging. This chapter explores types of interoperability in distributed systems in Section 2, and Section 3 lists the specific to healthcare. Section 4 talks about the benefits of semantic interoperability, and Section 5 enumerates the challenges to achieve the same. Section 6 lists some use cases for interoperability, and Section 7 concludes the chapter.

2 Types of interoperability

In general, there are three different types of interoperability—syntactic, semantic, and organizational interoperability. Syntactic interoperability is when two or more devices use popular information formats (such as XML and SQL) and communication protocols to communicate with each other. Semantic interoperability is the ability to interpret meaningfully and accurately exchange data automatically to generate meaningful outputs as specified by both systems. The common technique used to achieve semantic interoperability is to refer to a common reference model for information exchange. Organizational interoperability includes social, political, and legal entities working together for a common interest and/or exchange of information.

Reference frameworks are useful to relate different types of interoperability and to compare concepts, principles, methods, standards, models, and tools in a certain domain of concern [3]. Some of the reference models are the LISI reference model, ATHENA interoperability framework (AIF), healthcare information and management systems society (HIMSS), European interoperability framework, etc.

The Levels of Information Systems Interoperability (LISI) Reference Model provides the common vocabulary and structure required to discuss interoperability between IT systems. It illustrates five levels of interoperability as shown in Table 1 [3].

The AIF offers a composite structure and related reference architecture to capture research elements and solutions to interoperability problems. The AIF provides a related methodological structure to describe the interoperability strategy from the choice to assess cooperation to the maintenance of the solution and the reference rules for adopting the reference architecture [4]. The framework is divided into three parts: conceptual integration, application integration, and technical integration. Conceptual integration (organizational interoperability) focuses on concepts, metamodels, languages, and model relationships. The framework defines a reference model for interoperability that provides a basis for various systemizing aspects of interoperability. Application integration (semantic interoperability) focuses on methodologies,

Table 1 Five levels of interoperability defined in LISI.

Level	Interoperability
Level 0	Isolated systems (manual extraction and integration of data)
Level 1	Connected interoperability in a peer-to-peer environment
Level 2	Functional interoperability in a distributed environment
Level 3	Domain-based interoperability in an integrated environment
Level 4	Enterprise-based interoperability in a universal environment

Reused from F.B. Vernadat, Technical, semantic and organizational issues of enterprise interoperability and networking, IFAC Proc. Vol. 42 (4) (2009) 728–733.

standards, and domain models. The framework describes methods that provide guidelines, principles, and patterns that can be used to solve problems of interoperability. Technical integration (technical interoperability) focuses on software development and execution environments. The framework defines a technical architecture that includes development tools and execution platforms for integrating processes, services, and information [3].

The European Interoperability Framework is a standard framework that describes three fundamental levels of interoperability, namely, technical, semantic, and organizational interoperability, as shown in Fig. 1 [3].

Healthcare Information and Management Systems Society (HIMSS) classifies interoperability as foundational, syntactic, semantic, and organizational. Foundational interoperability defines the building blocks of information exchange between different systems by establishing the connection requirements required for one system or application to share and receive information from another system. Structural interoperability specifies the structure or format of data exchange. The syntax of the data exchange is defined by structural interoperability. It ensures that data exchanges can be interpreted at the level of the data field between information systems. Semantic interoperability is the ability of two or more systems to understand, exchange, and

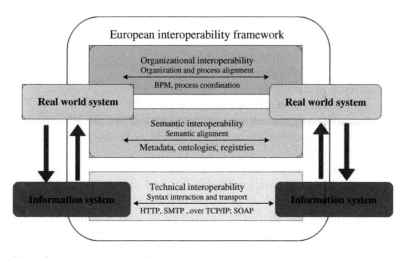

Fig. 1 European interoperability framework Redrawn from F.B. Vernadat, Technical, semantic and organizational issues of enterprise interoperability and networking. IFAC Proc. Vol. 42 (4) (2009) 728–733.

use information. Semantic interoperability makes use of the structuring of the data exchange and the codification of the data, including standards, publicly available vocabulary so that the receiving information management systems can understand the data. Organizational interoperability contains the technical components, policy, social, and organizational elements. These elements promote the secure, seamless, and timely communication and use of information within and between organizations and individuals [5].

3 Interoperability in healthcare

According to HIMSS, interoperability is the ability of different information systems, devices, or applications to connect, in a coordinated manner, within and across organizational barriers to access, exchange, and cooperatively utilize data among stakeholders, with the goal of optimizing the health of individuals and populations [5]. Healthcare systems demand interoperability among medical experts, clinicians, systems, and workflow. Domain-specific interoperability standards are needed and have to support cross-domain use. Communication and collaboration among healthcare systems and their components require the following [6]:
- openness
- scalability
- flexibility
- portability
- distribution at internet level
- standard conformance
- service-oriented semantic interoperability
- appropriate security and privacy services

For achieving the aforementioned characteristic HIS components, their relationship and functionalities should meet the following criteria:
- distributed architecture;
- component orientation (flexibility and scalability);
- model-driven and service-oriented design;
- separation of platform-independent and platform-specific modeling and separation of logical and technological views (portability);
- specification of reference and domain models at metalevel (semantic interoperability);
- interoperability at service level (concepts, contexts, and knowledge);

- common terminologies and ontology;
- semantic interoperability;
- security, safety, and privacy services;
- deep learning and neural networks (algorithms) [7].

Semantic interoperability must be incorporated with the health information system to get secure and high-quality healthcare. Semantic interoperability helps in sharing information between different stakeholders in the healthcare system. It helps in creating a common vocabulary that creates a way for accurate and reliable communication among computers. This communication between systems depends on the ability of different HIT systems to map different terms to shared semantics, or meaning. Semantic interoperability is viewed as critical for several healthcare initiatives like quality improvement programs, population health management, and data warehousing. Critical security issues of interoperability are whom to share, how much to share, and how to share such that no unauthorized access can be made to any data. Security can be ensured by using technologies like blockchain architectures to develop secured interoperable EHR systems [8, 9].

Semantic interoperability allows healthcare organizations to move from legacy systems. Heterogeneous HISs in healthcare environments use openEHR archetypes to represent the information contained in clinical EHR extracts and support the interoperability [10]. Table 2 compares the existing healthcare interoperability standards.

One of the major faults with the standards is that they are domain dependent and they are not built to include information of various formats from wearable devices. Moreover, what is achieved in the standards is primarily a way to attain syntactic interoperability, whereas a greater need is for semantic interoperability.

4 Use case of semantic interoperability in healthcare

A typical use case depicting the need for semantic interoperability is listed in the succeeding text. Patient A will undertake treatment in Hospital A for some medical condition and may need to move to Hospital B for some other issue. In the course of his life as he visits different hospitals, he manages to create a lot of data in the form of electronic health records scattered across various hospitals, as shown in Fig. 2. The summary of the treatment may be given to the patient at the time of discharge. But this is only a piece

Table 2 Comparison of existing interoperable standards in healthcare.

Properties	HL7	CCR	EN/ISO 13606	openEHR	IHE
Unified process modeling	Yes	No	No	No	No
Business modeling	Yes	No	No	No	Partial
Service-oriented process	Partial	No	No	No	Partial
Reference information model	Yes	No	Yes	Yes	Yes
Metamodel	Partial	No	No	No	No
Concept representation	Yes	Partial	Yes	Yes	No
Domain independent	No	No	No	No	No
Ontology driven	No	No	Yes	Yes	No
Vocabulary	Yes	No	No	No	No
Reference to terminology	Yes	Yes	Yes	Yes	Yes
Communication security	No	Yes	No	No	Yes
Inclusion of medical devices	Yes	No	Feasible	No	Feasible
Implemented	Yes	Yes	No	Partial	Yes
Commercial product	Yes	Yes	No	No	Yes

of limited information when compared with other details like nurse's notes and doctor's notes that might be captured during the stay.

If the EHRs are semantically interoperable, then data from different health providers can be consolidated automatically. This can be used by an intelligent clinical decision support system (CDSS) for intelligent summarization and giving alerts. When patients visit specialists, the system can check to see if there are any interdependencies with other treatments he has undertaken and give appropriate warning. If, in his medical history, there is an indicator of strong reaction to some drugs, this can be extracted and informed, thus ensuring better healthcare experience.

Another use case is the generation of a patient-centric, patient-mediated EHR system. EHRs of the patient from different sources can be consolidated and stored in the form of PHR. The patient can provide access to this information to those who need to know. In case of emergency, this information will be quickly accessible and will be useful to make informed decisions by caregivers. Currently, EHR software providers are using different standards that make the concept of interoperability challenges.

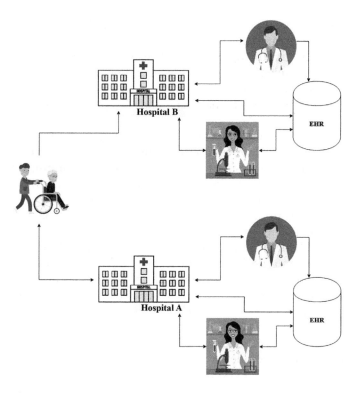

Fig. 2 Scenario where same patient visit multiple hospitals.

Another scenario is the requirement of different stakeholders having access to the data. Healthcare providers/administration/insurance all have different uses of these data. EHR is confidential information as far as the patient is concerned. So, semantic understanding of data can help in determining how data can be shared among different stakeholders in the best interest of the patient.

5 Benefits of semantic interoperability

The primary goal of the healthcare system is to provide better healthcare and safety to patients. Semantic interoperability allows the healthcare system to work together and helps to provide better patient care and effective healthcare delivery. In addition to the improvement of patient care, semantic interoperability offers the following benefits:

- Meaningful analysis: Semantic interoperability allows generating useful analytics about doctors' or clinicians' processes. It helps in identifying excesses of resources and thereby allows planning releases for clinicians for other critical work to boost productivity. The analytics will enable administrators for streamlining work balance and keeping track of staff schedules.
- Clinical process reengineering: Semantic interoperability enables clinical procedures to be reengineered to make it more efficient. When systems are fully interoperable, administrators can recognize excursions from account, when and how a clinician or physician behaves in an outside protocol.
- Resources utilization: Semantic interoperability analyzes and optimizes the use of resources.
- Clinical quality monitoring: Semantic interoperability helps in efficient monitoring of clinical processes and thereby provides better patient care. Semantic interoperability analytics help administrators to detect and avoid issues before they become problems.
- Solving issues in legacy systems: In the case of legacy systems, semantic interoperability allows legacy systems to operate in maintenance mode.

6 Challenges in semantic interoperability

Healthcare will move to a new dimension if semantic interoperability between EHR can be achieved, and the same can be used to derive intelligence. It can be given as an input to the clinical decision support system (CDSS), workflow management, and evidence-based healthcare. Some of the issues that need to be addressed to achieve semantic interoperability to share EHR among different stakeholders are as follows [1, 3]:

- Partial mapping of multiple sources: Combining the attributes identified in different systems leads to semantic differences, which lead to partial data mapping. This is due to inconsistent and nonformalized information structuring.
- Need for user intervention: It is challenging to identify common meaning and use of conflicting terms in a health information system without user intervention.
- Setting standards and guidelines: There is a need to construct a set of principles that ensure that updated rules are syntactically and (or) semantically correct, it is important to examine and define policy conflicts.

- Addressing contextual constraints: Currently, there are many vocabulary lists that are built, but identifying the right interpretation of similar concepts with different meanings remains a challenge.
- Existence of semantic difference in attributes: The existence of semantic difference in attributes and many times the meaning has to be inferred utilizing reasoning rules that form the foundation for the valid identification of similarities between them.
- Platforms for semantic interoperability: There is a need to leverage information retrieval techniques, neural networks, and artificial intelligence to assess the similarity of different elements in the profile.
- Medical terminologies: A bigger challenge is the correct understanding and interpretation of medical terminology.

Literature review done on the existing works revealed that there is a scope of much work in the limitations listed earlier. The findings of the literature survey are summarized in Table 3.

7 Research directions

Use of IoT devices like a wireless sensor and smartphone helps to achieve patient-centric health records (PHR). Semantic interoperability plays a crucial role in achieving patient-centric care. EHR, wireless devices, and information from healthcare environments need to be consolidated to create advanced CDSS. This CDSS must make real-time decisions on incoming EHR from personal health record (PHR) and historical EHR. The decision made must be based on clinical practice guidelines (CPG), validated by a domain expert. Advanced techniques like artificial intelligence, big data analytics, data mining, and cloud computing need to be used to improve decisions. Security and privacy of patients' EHR, network devices (wired and wireless) is another direction to improve healthcare.

8 Conclusion

Heterogeneous healthcare data are being generated from various sources like hospitals, laboratories, doctor prescription, nurse's notes, and wearable devices. This information is valuable for the construction of EHR. EHRs from various sources need to be interoperable so as to leverage the maximum benefit and provide the best care for the patients. Existing software are not designed with interoperability in mind. Many of the software do not follow any standard, and those that follow use different standards like

Table 3 Current works in semantic interoperability and its limitations.

Paper referred	Interoperability standards	Achieving semantic interoperability	Methodology	Purpose	Advantages	Limitations
[11]	HL7	Nil	Rational unified process, SOA, model-driven architecture (MDA), ISO 10746, generic component mod	Develop a framework for semantically interoperable HIS	Generic component model, not restricted to single standards	HL7 specifications have been partially reused
[6]	HL7 Version 3	HL7RIM	Model-driven, SOA	Highly dedicated and distributed healthcare facilities must interact and collaborate in a semantically interoperable manner to fulfill the challenge of high-quality and effective care	Model-driven, service-oriented design	Difference in generic and domain-specific knowledge concepts
[12]	HL7 and FHIR	Ontology and medical standards	OWL2 ontology	To provide monitoring of T1D patients	Distributed, semantically intelligent, cloud and mobile technologies	Uncertain nature of medical Data not handled

[10]	openEHR	Archetype and ontology	openEHR and agent-based technology	Developing e-health applications and reusing legacy systems	Use of agent-based system and archetype	Questionnaire used in assessment is not entirely subjective, as each participant can respond to it in a specific manner. Health professionals grouped and their vast expertise, findings have been generalized
[13]	HL7	Ontology	Nil	Semantic interoperability requires different levels of inter- and intraorganizational integration and is challenging	Nil	Nil
[14]	Nil	SNOMED-CT, ontology	Service-oriented architecture(SOA)	Reuse of CDSS by encapsulating in web service	Reuse of CDSS	Political, legal, organizational, and technical challenges
[15]	openEHR ISO EN 13606	Ontology, archetype	Model-driven architecture	Semantic interoperability of two EHR standards: openEHR and ISO EN 13606	Different interpretations for different purposes	Only two standards compared
[16]	Integrating the Healthcare Enterprise (IHE)	HL7 CDA	Nil	Secure P2P agent coordination framework	To enable secure interactions among communities	Nil

Continued

Table 3 Current works in semantic interoperability and its limitations—cont'd

Paper referred	Interoperability standards	Achieving semantic interoperability	Methodology	Purpose	Advantages	Limitations
[17]	HL7	HL7, ontology	Generic component model	Introduces the fundamental paradigms, specifications, architectural reference models, concepts, and formalization principles and processes for the development of comprehensive service-oriented personalized eHealth	PHR, pHealth	Key demand for personal health is to bridge disciplines including ontology
[18]	HL7	HL7 RIM, UML, SNOMED C T	Local information models	Build a digital electronic health record in the cardiology and dental medicine	Structured data entry over free-type data entry	Concept-related issues and their mapping of global classification schemes require close collaboration with physicians
[19]	HL7, CCR, openEHR, IHE, and EN/ISO 13606	Nil	Nil	Analysis and evaluation of EHR approaches using various standards	Nil	Nil
[20]	HL7 CDA, CEN 13606, and openEHR	Ontology	Unified fuzzy ontology	Framework for distributed EHR based on fuzzy ontology	Fuzzy ontology	Nil

HL7 and FIHR. Thus, to improve the healthcare system and provide better quality care, there is a need to provide syntactic and semantic interoperability among EHR. Semantic interoperability enables healthcare systems to work together and achieve better patient outcomes and effective healthcare. By achieving semantic interoperability and with artificial intelligence, big data analytics and data mining real-time decisions for better care can be made. The rich information that can be consolidated from the EHRs combined with artificial intelligence techniques can revolutionize modern healthcare delivery.

References

[1] ISO/TR 20514:2005, Health Informatics—Electronic Health Record—Definition, Scope and Context by ISO/TC 215, Multiple, Distributed through American National Standards Institute, 2007, pp. 1–27.

[2] J.C. Nieves, A. Espinoza, Y.K. Penya, M.O. De Mues, A. Pena, Intelligence distribution for data processing in smart grids: a semantic approach, Eng. Appl. Artif. Intell. 26 (8) (2013) 1841–1853.

[3] F.B. Vernadat, Technical, semantic and organizational issues of enterprise interoperability and networking, IFAC Proc. Vol. 42 (4) (2009) 728–733.

[4] A.-J. Berre, B. Elvesæter, N. Figay, C. Guglielmina, S.G. Johnsen, D. Karlsen, T. Knothe, S. Lippe, The ATHENA interoperability framework, in: Enterprise Interoperability II, Springer, London, 2007, pp. 569–580.

[5] HIMSS (Ed.), What Is Interoperability?, HIMSS, 2019 Accessed: September 11, 2019. https://www.himss.org/library/interoperability-standards/what-is-inter
operability.

[6] B.G.M.E. Blobel, K. Engel, P. Pharowe, Semantic interoperability, Methods Inf. Med. 45 (04) (2006) 343–353.

[7] Samui, P., Roy, S. S., & Balas, V. E. (Eds.). (2017). Handbook of Neural Computation. Academic Press

[8] Ekblaw, A., Azaria, A., Halamka, J. D., & Lippman, A. (2016, August). A case study for blockchain in healthcare:"MedRec" prototype for electronic health records and medical research data. In Proceedings of IEEE Open & Big Data Conference (vol. 13, p. 13).

[9] Martínez-Villaseñor, Ma Lourdes, Luis Miralles-Pechuan, and Miguel González-Mendoza. "Interoperability in electronic health records through the mediation of ubiquitous user model." In International Conference on Ubiquitous Computing and Ambient Intelligence, pp. 191–200. Springer, Cham, 2016.

[10] J.L.C. de Moraes, W.L. de Souza, L.F. Pires, A.F. do Prado, A methodology based on openEHR archetypes and software agents for developing e-health applications reusing legacy systems, Comput. Methods Prog. Biomed. 134 (2016) 267–287.

[11] D.M. Lopez, B.G. Blobel, A development framework for semantically interoperable health information systems, Int. J. Med. Inform. 78 (2) (2009) 83–103.

[12] S. El-Sappagh, F. Ali, A. Hendawi, J.H. Jang, K.S. Kwak, A mobile health monitoring-and-treatment system based on integration of the SSN sensor ontology and the HL7 FHIR standard, BMC Med. Inform. Decis. Mak. 19 (1) (2019) 97.

[13] S. Bhartiya, D. Mehrotra, A. Girdhar, Issues in achieving complete interoperability while sharing electronic health records, Procedia Comput. Sci. 78 (2016) 192–198.

[14] L. Marco-Ruiz, C. Pedrinaci, J.A. Maldonado, L. Panziera, R. Chen, J. Gustav Bellika, Publication, discovery and interoperability of clinical decision support systems: a linked data approach, J. Biomed. Inform. 62 (2016) 243–264.

[15] C. Martínez-Costa, M. Menárguez-Tortosa, J.T. Fernández-Breis, An approach for the semantic interoperability of ISO EN 13606 and OpenEHR archetypes, J. Biomed. Inform. 43 (5) (2010) 736–746.

[16] V. Urovi, A.C. Olivieri, A.B. de la Torre, S. Bromuri, N. Fornara, M. Schumacher, Secure P2P cross-community health record exchange in IHE compatible systems, Int. J. Artif. Intell. Tools 23 (01) (2014) 1440006.

[17] B. Blobel, Architectural approach to eHealth for enabling paradigm changes in health, Methods Inf. Med. 49 (02) (2010) 123–134.

[18] J. Zvárová, P. Hanzliček, M. Nagy, P. Přečkova, K. Zvára, L. Seidl, V. Bureš, D. Šubrt, T. Dostálová, M. Seydlová, Biomedical informatics research for individualized life-long shared healthcare, Biocybern. Biomed. Eng. 29 (2) (2009) 31–41.

[19] B.G.M.E. Blobel, P. Pharow, Analysis and evaluation of EHR approaches, Methods Inf. Med. 48 (02) (2009) 162–169.

[20] Adel, E., El-Sappagh, S., Barakat, S., & Elmogy, M. (2019). A unified fuzzy ontology for distributed electronic health record semantic interoperability. In U-Healthcare Monitoring Systems (pp.353-395). Academic Press.

Difficulty with language comprehension and arithmetic word problems due to hearing impairment: Analysis and a possible remedy through a new Android-based assistive technology

Sandipa Roy, Arpan Kumar Maiti, Gopal Krishna Basak, and Kuntal Ghosh
Indian Statistical Institute, Kolkata, India

1 Introduction

Persons with satisfactory levels of hearing ability may learn and acquire written languages as a visual representation of spoken language. However, such a communication system is not always appropriate and effective for every human being, especially for those who are suffering from speech and hearing insufficiency. It is evident that there is a significant communication gap between people of different hearing abilities and this gap is increasing with the availability of more and more modern amenities, which are often inaccessible to the speech and hearing impaired (HI).

In general (for normal hearing [NH] persons), consciously or unconsciously, a common audiovisual system is used for cognition and memorization in almost every circumstance. It is a neurological system in the brain that has been trained since childhood. However, this system does not work in the case of HI individuals. For these people, symbols and sounds are not being properly corresponded since all letters or words are encoded with phonemes.

These letters or words are just like symbols without any defined meaning for HI persons.

Hearing impairment is a common feature for speech and HI people. There is a huge gap in communication and consequently development between NH and HI persons. An attempt is being made here to reduce this gap between these communities to the minimum possible level, and to make communication between the two groups more effective. HI or mute persons are connected with the outer world with the help of gestures, postures, facial expressions, sign language, or some other means. These modes of communication have been their preferred mode of interaction with society in their day-to-day lives. Under this backdrop, we make a modest attempt to establish a system and create a common platform in order to facilitate communication between NH and HI individuals.

The All India Federation of the Deaf estimates that there are around 4 million mute and more than 10 million HI persons in India. Studies have revealed that one out of every five deaf people in the world is from India. More than 1 million deaf adults and around 0.5 million deaf children in India use Indian Sign Language (ISL) as a mode of communication. ISL is not only used by deaf people but also by the hearing parents of deaf children, the hearing children of deaf adults, and hearing educators of HI people [1].

The proposed work is expected to help HI people by providing an option for recognizing symbols in the absence of sound through some visual clues. It is proposed to:

1. reduce the communication gap between HI and NH persons. The objective is to support the integration of HI people into hearing society and facilitate a common, comfortable platform in the public sector.
2. develop writing skills to a certain extent by increasing the stock of words or vocabulary and mapping it into semantic connotation via sign inputs.
3. develop a signer independent platform to improve communication via signs for purposes of multiple language acquisition. In the proposed method two iconic strategies are being followed:
   - holding or grasping an imagined object in action using the hand; and
   - measuring the shape or dimension of an imagined object by hand representation.

With the help of these strategies, we explore cognitive association of real-world objects using ISL. The proposed method of ISL follows both of the aforementioned strategies to create a cognitive basis for differentiating object from action (for HI persons). As such, HI children would come to know more about the vocabulary, and their stock of words would increase automatically in the process. The proposed method also helps by reducing the communication gap between HI and NH persons. Thus, HI persons would not only become educated, but in the process they would also be able to communicate with NH persons more easily and effectively.

No magical formula exists in this universe for ensuring effective verbal communication since it is too complicated a method. However, the communication process, especially for HI persons, can be improved. There is obviously a significant communication gap between HI persons and NH persons. HI persons have great difficulties with respect to reading and writing.

Visits were conducted at the Calcutta Deaf and Dumb School, Ali Yavar Jung National Institute of Speech and Hearing Disabilities (AYJNISHD), Eastern Regional Center, Kolkata, Indian Institute of Psychometry (IIP), Kolkata, and Center for Development of Advanced Computing (CDAC), Kolkata. The views and opinions were collected through interaction with parents, teachers, principals, and many others associated with HI or mute persons. It was noted that existing oral communication, lip movements, writing methods, and speech therapy were employed in communication in a scattered way. The HI individuals encountered during our visits displayed poor writing skills; they just put down words one after another to form sentences without any proper meaning. They found it difficult to understand or realize any words and their meaning. They could not understand the meaning of any passage properly. There is also a dearth of research work that deals with approaches to proper sentence making, vocabulary power, and mapping to semantic connotations via sign inputs on Indian Sign Language Recognition (ISLR) for the speech and HI. It was also found that a systematic thought process and ideas are not properly developed by the applications of existing methods. With this understanding, we divided our work into two main parts (part A and part B). Part A dedicated for statistical analysis toward understanding the problem and part B is dedicated to possible solution through app-based technology are described in given below.

A Statistical analysis toward understanding the problem

First, we describe the problem in Section 2, present the survey methodology in Section 3, and statistically analyze the survey outcomes in Section 4.

B Possible solution through app-based technology

Second, we discuss the app's requirements in Section 5, present its design in Section 6, and examine results in Section 7.

A Statistical analysis toward understanding the problem

2 Description of the problem

Most of the information that comes to our brain, comes through our sensory organs. The eyes and ears are the most significant parts of this process (Fig. 1). Then the brain processes the received information to extract meaning out of it and helps us understand our environment. Some of the information is sent to our memory to be stored for either the short term or long term. These stored memories develop our thought processes and in turn shape our personality and behavior. The immediate output of the perceived information is in the form of communicative speech or gesture. The concepts that we understand are well mapped in our brains. In the event of hearing impairment, the brain receives only partial information and all the information is not properly mapped into the brain (Fig. 2). In this case, the eye is the only major sensory organ that sends information and hence the dependence of the brain is on the eye to receive information. The brain receives only video signals and no audio signals. Without audio signals vision does not corresponded to an appropriate meaning of a written text, and hence a symbol-text pair cannot be properly mapped in the brain. The brain fails to recognize the text from the symbol and vice versa. In absence of hearing capability, any auxiliary information has to be provided through vision. Hence as a substitute to the audio signal we give some hand gesture to provide more information to the brain. To make the brain recognize any word, we

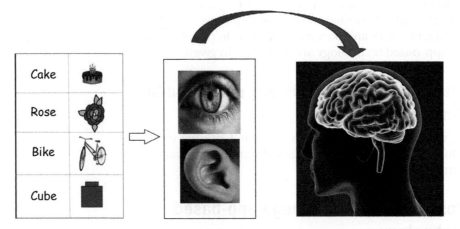

Fig. 1 The brain of an NH person recognizes the symbol through sound with vision.

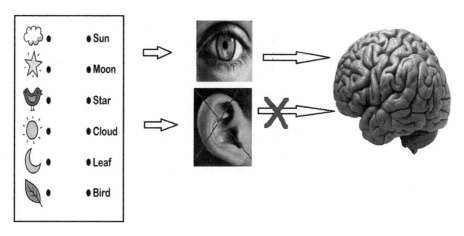

Fig. 2 Symbols are not properly recognized due to hearing impairment.

associate a gesture with it so that the text word is mapped with the corresponding symbol in the brain (Fig. 3). If we are able to introduce a sign for every word in the vocabulary, then perhaps there will be much less language-oriented problems for HI persons.

2.1 Statement of the problem and verification

Speech-impaired and HI students have problems with language. Despite having knowledge of basic calculations in mathematics, HI persons do not perform well with word problems or comprehensive sums problems.

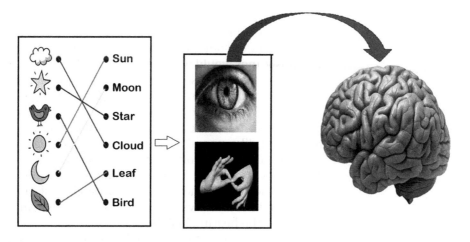

Fig. 3 The brain recognizes the symbol through sign.

2.2 Hypothesis and basic assumptions

The main hypothesis and related assumptions are as follows.
- HI students are good in basic mathematical calculations like NH students (addition, subtraction, multiplication, and division).
- When a mathematical problem is given through a comprehension format, HI students' performance drops due to poor language understanding. This shows that they are unable to comprehend the given language.
- Hearing impairment is the root cause of such language inefficiency may be in HI students.
- Due to hearing impairment, HI students face problems in language, like with sentence formation, vocabulary understanding, and so on.

2.3 Verification of the actual scenario through first-hand survey

We conducted a survey to test proficiency in arithmetic as well as language.
- To test their ability of doing arithmetic, we gave HI students some basic calculations like addition, subtraction, multiplication, division, and simple summation to measure their reasoning capability. Further, we constructed some story-type mathematical problems to specially analyze their comprehensive arithmetic solving capability. For example: there are 18 cake packets in a shop. The shopkeeper sales 8 cake packets, and after that, he collects 10 more cake packets for sale. How many cake packets are now available in the shop?
- To identify their language understanding problem, we gave HI students a comprehension-like paragraph followed by questions. If they did not understand the meaning of the paragraph obviously, they would not be able to give the proper answer. To identify the root cause of their language problems, we gave them jumbled-up words to verify their sentence construction capability. We also provided some special words for filling up the blanks in sentences. These special words resemble, for an instant, like, sound of raindrop (pitter-patter), the sound of wall clock (ding-dong). In such cases, NH child pick up the words from the environment along with the associated sound made by NH adult. Hence through such an example, one would be able to verify if the language problem is directly related to the hearing inability.
- We verified whether the HI individuals had actual difficulty in learning language from written language due to their hearing inability.

3 Survey methodology

3.1 Sample

The initial survey was conducted for speech-impaired and HI learners from a few Bengali medium schools in and near Kolkata to assess their particular needs and the difficulties they face in their learning. The subjects of the study consist of 98 students from such schools, out of which 48 were from Bengali medium schools meant for the speech and hearing impaired students and 50 were from regular Bengali medium school for NH students from the outskirts of the city of Kolkata. Out of the 48 HI students 23 were female and 25 were male students. All the 48 students were 100% hearing disabled based on PwD Act, 1995. There were 50 NH students of the same age in the control group. Both groups were associated with the same socioeconomic strata varying from lower-middle to the upper-middle-class background.

3.2 Ethics approval

We obtained ethical permission from schools as well as parents of the specific students. They were assured that all information would be safe and used only for research purposes.

3.3 Survey topic: Arithmetic and language proficiency test

An arithmetic and language proficiency test was conducted for HI students as well as NH students. We divided our questions into two parts: one for language and the other for arithmetic proficiency.

3.3.1 Arithmetic proficiency test

We divided the arithmetic proficiency test into two groups.
- *Numerical ability*: Basic arithmetic calculations (addition, subtraction, multiplication, division, and simplification sums) were given to verify the logic sense.
- *Comprehensive arithmetic*: Language-based or story-based arithmetic problems with similar simple arithmetic calculations were given to verify language understanding capability.

3.3.2 Language proficiency test

To find out the root cause of the language-related problem we focused on these following areas and compared the HI children to the NH children.

- Complete sentence formation capability: The students were given a set of jumbled words and asked to arrange the words to make a proper sentence.
- Choose the correct pronoun-specific verb form in daily communication: They were tested if they were aware of the simple words used in daily communication. These are words that an NH child picks up from her/his environment.
- Choose the correct word that resembles the sound (onomatopoeia).
- Comprehension: To identify if they have any language problem, we gave both groups a comprehension passage and asked them to select correct answers from a set of options. If they understood the passage then they would be able to answer.

3.4 Object to verify

Two objects are fixed for verification.
- Whether the HI group has genuine issues in understanding arithmetic word problems resulting from their language deficiency.
- Whether the HI group has genuine problems in learning language due to their hearing deficiency.

3.5 Methodology

We provided the same question paper (arithmetic and language proficiency test) to both groups. Marks of each and every question are calculated in percentages.

1 represents the correct answer.
0 represents the incorrect answer.

3.6 Outcomes of the survey

The outcomes of the arithmetic proficiency test and language proficiency test are presented in the sections that follow.

3.6.1 Arithmetic proficiency test

From Fig. 4, we can see that both HI and NH students are good at basic mathematical calculations like addition, subtraction, multiplication, and division. We see that HI students perform well in basic calculations, but perform poorly when given a similar problem in comprehension format due to their poor language understanding (Fig. 5). This shows that HI students are unable to comprehend the given language. We confirm their language problem from Tables 2–4.

Chapter 3 Difficulty in language comprehension and arithmetic word problems **37**

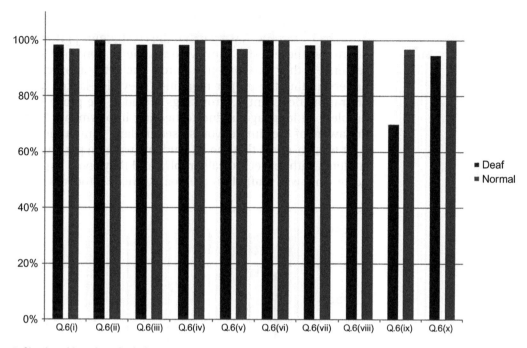

Fig. 4 Simple arithmetic calculations.

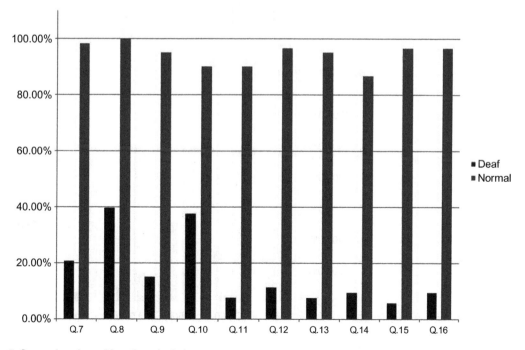

Fig. 5 Comprehensive arithmetic calculations.

3.6.2 Language proficiency test

From Figs. 6 and 7, we see that the HI students performed very poorly on these tests. They were unable to construct complete sentences; they just put the words one after another without understanding the proper meaning. They also lacked any sense of grammatical syntax. From Figs. 8 and 9, we prove that their problem in language is due to hearing impairment. They were not aware of the simple words used in daily communication.

It was observed that the HI students were on par with the NH students with respect to calligraphy, and they could strongly replicate what was drafted on the board. The HI students could not

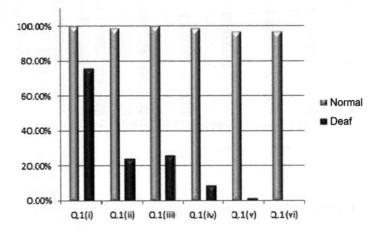

Fig. 6 Complete sentence formation.

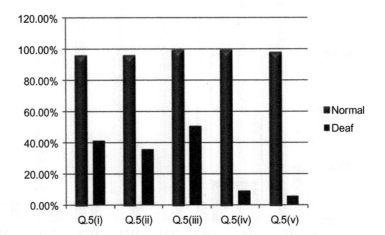

Fig. 7 Comprehension.

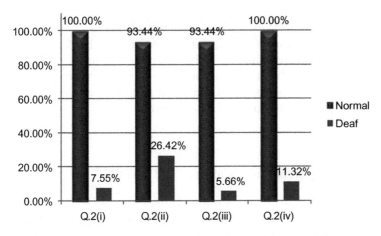

Fig. 8 Select the correct word that resembles the sound (onmatopia).

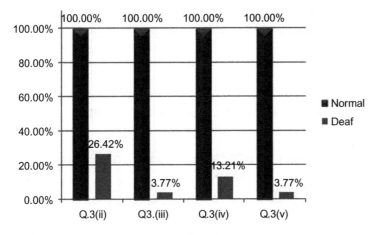

Fig. 9 Select the correct pronoun-specific verb.

generate a self-thought answer from their vocabulary. It was distinguished that they were unable to follow the grammatical sense of language, majorly concerning their proper sentence configurations, usage of tenses, and perception of synonyms and antonyms. They lacked in arithmetic, mensuration, and so on where language knowledge is exceedingly required. They could not discriminate similar words and contrasting words. As a consequence, they were not interested to read newspapers, storybooks, or other materials, which further diminished their ability to learn and thus improve their understanding.

4 Statistical analysis of the survey outcome

For each of the following statistical analyses, the data of which are represented in Tables 1–4, the null hypothesis and the alternative hypothesis both remain the same as stated below:

H0 (null hypothesis): No change or gap between the treatment and the control group (i.e., $\mu_1 = \mu_2$).

Table 1 Simple arithmetic calculations NH with HI.

	Q.6(i)	Q.6(ii)	Q.6(iii)	Q.6(iv)	Q.6(v)	Q.6(vi)	Q.6(vii)	Q.6(viii)	Q.6(ix)	Q.6(x)
NH	96.72%	98.36%	98.36%	100.00%	96.72%	100.00%	100.00%	100.00%	96.72%	100.00%
HI	98.11%	100.00%	98.11%	98.11%	100.00%	100.00%	98.11%	98.11%	69.81%	94.34%
Test_statistics	−0.47213	−1.0083	0.099893	1.00957	−1.43798	0		1.00957	1.00957	4.013153	1.78325545
P_value	.159209	.078328	.230107	.078175	.03761	.25	.078175	.078175	1.5E−05	.01863618

Table 2 Comprehensive arithmetic calculation NH with HI.

	Q.7	Q.8	Q.9	Q.10	Q.11	Q.12	Q.13	Q.14	Q.15	Q.16
NH	98.36%	100.00%	95.08%	90.16%	90.16%	96.72%	95.08%	86.89%	96.72%	96.72%
HI	20.75%	39.62%	15.09%	37.74%	7.55%	11.32%	7.55%	9.43%	5.66%	9.43%
Test_statistics	13.37324	8.986763	14.1739	6.833055	15.69632	17.38152	19.17895	13.1291	23.29991	18.90446
P_value	0	0	0	2.08E−12	0	0	0	0	0	0

Table 3 Complete sentence formation NH with HI.

	Q.1(i)	Q.1(ii)	Q.1(iii)	Q.1(iv)	Q.1(v)	Q.1(vi)
NH	100.00%	98.36%	100.00%	98.36%	96.72%	96.72%
HI	75.47%	24.53%	26.42%	9.43%	1.89%	0.00%
Test_statistics	4.150301	12.04529	12.15084	20.52903	32.16765	42.42051
P_value	8.3E−06	0	0	0	0	0

Table 4 Comprehension NH with HI.

	Q.5(i)	Q.5(ii)	Q.5(iii)	Q.5(iv)	Q.5(v)
NH	96.72%	96.72%	100.00%	100.00%	98.36%
HI	41.51%	35.85%	50.94%	9.43%	5.66%
Test_statistics	7.73058	8.73263	7.144021	22.5566	25.99307
P_value	2.66E−15	0	2.27E−13	0	0

H1 (alternative hypothesis): Difference exists between the two groups (i.e., $\mu_1 \neq \mu_2$).

The questions have been grouped differently as reflected by the four tables, in order to gain a deeper insight into the problem. Let us come to these analyses one by one.

4.1 Statistical data analysis: Simple arithmetic calculations (Table 1)

Hypothesis of equality of ability (or other indicators) between two groups would be rejected if the *P*-value is less than .05. (*H1*: $\mu_1 \neq \mu_2$) with $\alpha = 0.05$.

Comment: Since all the *P*-values for Q.6(i)–Q.6(viii) and Q.6(x) are greater than .05, we fail to reject our null hypothesis (i.e., accept the null hypothesis). Since most of the *P*-values with respect to question number 6, as provided in Table 1, are greater than .05, we fail to reject our null hypothesis at 5% level of significance. In other words, there exists no difference between the treatment and control groups in this respect, and any variation observed may be explained through sampling variation. Interestingly, however, there exists one or two such questions (like question Q.6(x)) for which the *P*-value is less than .05. These, we note, are the simplification sums where one needs to understand and apply the Bracket-Of-Division (BODMAS in the proper order rule). For these types of problems, therefore, we reject the null hypothesis at 5% level of significance, implying that sampling variation alone cannot account for the difference between the two groups. It may be linked to the requirement of comprehension of some verbal instructions for solving such sums, which the treatment group may be lacking. All the calculations in Tables 1–4 are done assuming that the given sample sizes are large enough so that the test statistics can be assumed to be normally distributed and the P-values may be found from the normal table.

4.2 Statistical data analysis: Comprehensive arithmetic calculations (Table 2)

Since the P-value is negligible (even less than .001) ($H1$: $\mu_1 \neq \mu_2$) hypothesis of equality of ability doing comprehensive arithmetic word problems (or other indicators) between two groups are rejected in favor of the alternative hypothesis $H1$.

Comments: The results indicate that the difference of ability between the two groups is very significant.

4.3 Statistical data analysis: Complete sentence formation (Table 3)

Since the P-value is negligible (even less than .001), hypothesis of equality of ability of doing complete sentence formation (or other indicators) between the two groups is rejected in favor of alternative hypothesis $H1$.

Comment: The results indicate that the difference of ability between the two groups is very strongly significant.

4.4 Statistical data analysis: Comprehension (Table 4)

Since the P-value is negligible (even less than .001), hypothesis of equality of ability (or other indicators) between the two groups is rejected in favor of alternative hypothesis $H1$.

Comment: The results indicate that the difference of ability in comprehension between the two groups is very significant.

Overall comment

From the statistical analysis presented, we felt the urgent need for some assistive technology as a probable solution to resolve the language as well as comprehensive arithmetic problems. Through an Android-based mobile app, we made an attempt to improve HI students' vocabulary knowledge simultaneously with their arithmetic problem-solving capabilities. For vocabulary knowledge acquirement (VKA), we adopted the strategy of using antonyms.

This remedial approach constitutes the second part of our work, which we describe in the following three sections.

B Possible solution through app-based technology

5 App requirement analysis

Through our survey, we learned that HI children put primary words one after another without constructing a complete sentence according to proper grammatical syntax. They can replicate written prose from the blackboards and books without comprehending the meaning. They also lack the ability to express their self-thought responses in proper written format. They cannot form proper sentences due to lack of understanding of usage of tense, and it is hard for them to find the relevant words as per the situation due to insufficient vocabulary (e.g., synonyms and antonyms and their appropriate applications) [2]. These inefficiencies lead to their lack of interest in reading books and newspapers, which further affects their reading comprehension ability. We have also seen in our survey (Figs. 4 and 5) that they are proficient in basic arithmetic calculations like addition, multiplication, and division, which implies basic logical sense (Fig. 4). However, when faced with story-based arithmetic problems involving similar basic calculations, HI students cannot comprehend, due to lack of vocabulary knowledge. Additionally, they were also found severely lacking in mathematical theorems, comprehensive mathematics like problem sums, mensuration, and so on where language understanding is an essential part of solving the problems. The aforementioned issues motivated us in designing the app aimed at teaching HI students to solve language-based arithmetic problems. This results in better language comprehension through understanding the arithmetical words with opposite meanings (like profit-loss, increase-decrease, variable-constant, etc.) [3]. The language levels of preschool children who are HI and hard of hearing are delayed, often 2–3 years, behind their hearing peers [4]. These children actually possess very small size thesauri as compare to their NH peers. This, in turn, further reduces their (HI) rate of learning new words [5]. If first language acquisition is delayed, it negatively affects the child's linguistic ability permanently [6]. In case of an HI child, the child's inability to process linguistic information through listening decreases his or her academic performance drastically [7].

It is also observed that HI children are much weaker in mathematics relative to their hearing peers. The two main reasons being their relatively weak linguistic skills and less social interaction due to inferior teacher expectations and inadequate teaching methods [8]. It is already established that the delay in their language acquisition directly affects their capacity to resolve arithmetic words problems and comprehensive arithmetic performance [9]. Their proficiency in mathematics including academic domains is less than that of same-aged NH children [10–12]. So to increase solving capability of comprehensive arithmetic, we first need to strengthen their vocabulary knowledge.

First, we identified that the breadth and depth in VKA is deeply integrated with both synonyms and semantically equivalent uses of words as well as antonyms and opposite uses of words. In the case of speech and HI children, providing the knowledge of antonyms increases their reading comprehension considerably. Studying antonyms is a valuable tool for VKA (beyond vocabulary size). It is already proved that the study of antonyms has a massive impact on American Sign Language. The learning of antonyms teaches one the relationships among words and thus helps increase the breadth and depth of vocabulary knowledge. Clearly, one with a better knowledge of antonyms has a larger vocabulary [13]. Breadth and depth of vocabulary knowledge is directly proportional to the usage of antonyms. It has been found that the recognition of antonyms occurs through a language-enriched right hemisphere [14]. Thus, breadth of vocabulary helps in developing the building blocks required for complicated language skills and higher-order thinking skills [15, 16]. Vocabulary size is bigger for those people who know more antonyms. The information of opposites represented by the antonyms that demonstrate the relationship between two words elaborates the meaning of each individual word and thus improves vocabulary depth [13]. Children have unique and creative ways of using antonyms intuitively from an early age [17–19]. They can comprehend antonyms even before the age of 4 years [20]. Due to the delay in their early intervention [21], speech-impaired and HI children have delayed experience of language comprehension [4].

Presently this research is being implemented in the Indian Statistical Institute, an organization that has made numerous contributions to society through mathematical and statistical analysis. So, we have specially emphasized mathematical skill development for HI children, and for that purpose we are designing an app to solve their mathematical word problems with VKA. Initially, we restricted ourselves to creating a database of vocabulary on the basis of books of the West Bengal Board of Secondary Education from class I to VIII. The vocabulary is chosen based on the feedback from the teachers or special educators who teach HI students.

Some of the key features of our proposed work (mobile app) at the present stage of design and development are:
- ISL dictionary, which includes 11 Indian provincial languages with special emphasis on Bengali language. We are trying to describe the differential meaning of Bengali synonyms-antonyms through simple sentences.
- Help HI people to communicate more with NH people.
- Offer a valuable tool for parents to teach their HI or mute child.
- Help to improve comprehension ability and logical sense of the word through synonyms and antonyms and sentence formation example.
- General antonym study (e.g., after-before, close-far) to improve depth of vocabulary.
- Mathematical antonym study (e.g., constant-variable, addition-subtraction, proportional-inversely proportional) to improve language as well mathematical conception.
- Help to improve mathematical reasoning and the ability to solve arithmetic.

6 App design

User interface (UI) design is the most important part of this app. UI is suitably designed maintaining the following characteristics:
- Since India is a country with high linguistic diversity, we have used multiple languages including English so that this application is accessible to everybody, irrespective of their region. This app can be used as a dictionary for ISL in any of the 11 regional languages including Nepali and Bengali, which are also the official languages of the neighboring countries of Nepal and Bangladesh, respectively. The app includes English also to encourage the higher education of users.
- We have taken care to develop the vocabulary for better understanding ability as well as semantic advancement. We have included sign video, symbols, antonyms, and synonyms for utilization of these words in various sentences, which will improve user aptitude. Application of the same word in multiple sentences is explained adequately. To enable HI users to understand the semantic contrast between lexicons, we have taken examples of synonyms along with different example of antonyms, and included both for understanding the contrast. For example, White: the color of milk is white; Black: the color of hair is black.
- A powerful search engine has been designed and provided for searching words from a considerably large expandable database.

- Antonyms are carefully included in the dictionary to build up the semantics and conceptualize the topic. The vocabulary is specially chosen from schoolbooks in vernacular languages so that comprehension and problem-solving aptitude can be enhanced. We have painstakingly created the vocabulary for cognitive as well as linguistic understanding. Mathematical words have their related words to strengthen logical reasoning. These words are also taken from the arithmetic books used in regional language schools (nursery to class VIII) for improving their perception and mathematical capabilities, for example, variable-constant, maximum-minimum, addition-subtraction, and so on. Some related contrasting words in mathematics like square-rectangle and circle-oval are carefully included in our app.
- The app can be used as an ISL dictionary in 11 local languages spoken in the country. Preliminary, we concentrated on the Bengali lexicon with synonyms-antonyms along with different syntactic patterns. Since our introductory work toward promoting the Android ISL app has been started from the state of West Bengal, Bengali being the L2 against ISL as L1, we decided to take the help of a Bengali sign language corpus enriched by India's neighboring SAARC community, the Bangla Ishara Bhasaha Avidhan or Bengali sign language dictionary composed by Bangladesh Sign Language Committee, 1997 edition [22]. Fig. 10 represents a layout synopsis of the designed app.
- Around 2000 words (separately offered in 11 languages) have been signed and made available as part of the app.

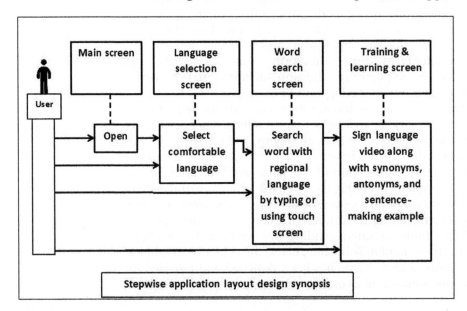

Fig. 10 Layout design of the app.

6.1 Steps for database creation

As recommended by special educators, cognitive scientists, mathematicians, and sign language experts, more than 2000 words (frequently used in textbooks, including mathematics) were chosen from Bengali books of the West Bengal Board of Secondary Education (classes II–VIII) for the dictionary and translated into 11 different languages using the help of Google Translator (English, Bengali, Malayalam, Tamil, Telugu, Marathi, Punjabi, Gujarati, Nepali, Hindi, and Urdu). Initially, we are promoting the Bengali vocabulary app-based database with synonyms and antonyms (of adjectives, adverbs, verbs) with usage in sentences. This kind of database will be part of this app for different languages in the future as several language specialists from different regions are needed to complete the app. For other languages except for Bengali at present, only the words and sign videos are available in the respective languages. In Fig. 11, we present a snapshot of the overall database.

6.2 Video recording

ISL videos have been collected for a sign language database. All the sign videos were developed and validated by the Indian Sign Language Research Training Center (ISLRTC), an autonomous body under the administrative and financial control of the Department of Empowerment of Persons with Disabilities (Divyangjan), Ministry of Social Justice and Empowerment, Government of India.

6.3 App development

The intended application was performed on the Android operating system, an easy-to-use operating system used in mobile accessories such as smartphones and tablet computers. The

serialNo	englishWord	englishToHindi	englishToBengali	englishToNepali	englishToPunjabi
1	A	A----एक	A----একজন	A----एक	A----ਇੱਕ
2	Above	Above----पर	Above----উপরে	Above----माथि	Above----ਉੱਪਰ
3	Absent	Absent----अनुपस्थित	Absent----অনুপস্থিত	Absent----अनुपस्थित	Absent----ਗੈਰਹਾਜ਼ਰ
4	Accept	Accept----लेना	Accept----গ্রহণ করা	Accept----स्वीकार	Accept----ਸਵੀਕਾਰ ਕਰੋ
5	Access	Access----पहुंच	Access----প্রবেশ	Access----पहुँच	Access----ਪਹੁੰਚ
6	Accident	Accident----दुर्घटना	Accident----দুর্ঘটনা	Accident----दुर्घटना	Accident----ਦੁਰਘਟਨਾ
7	Accuse	Accuse----आरोप	Accuse----অভিযুক্ত করা	Accuse----आरोप	Accuse----ਦੋਸ਼
8	Ache	Ache----दर्द	Ache----ব্যাথা	Ache----दुख्यो	Ache----ਦਰਦ
9	Active	Active----सक्रिय	Active----সক্রিয়	Active----सक्रिय	Active----ਸਰਗਰਮ
10	Actor	Actor----अभिनेता	Actor----অভিনেতা	Actor----अभिनेता	Actor----ਅਦਾਕਾਰ
11	Add	Add----योग	Add----যোগ	Add----थप	Add----ਜੋੜੋ
12	Address-1	Address-1----पता 1	Address-1----ঠিকানা 1	Address-1----ठेगाना-1	Address-1----ਪਤਾ 1

Fig. 11 A snapshot of the overall database.

prevalence of this operating system provides access to a broad range of beneficial libraries and tools that can be utilized to build elegant applications. This application is developed using the following software.

6.3.1 Android studio

Android Studio is a software to build the Android-based application that can be installed on a smartphone and tablet. It also has the facility to test the app using the simulated mobile app provided by this software.

6.3.2 SQL lite

This is the common widely deployed database engine that is accepted today by several browsers, operating systems, and embedded systems. This database is used to save the sign video and vocabulary in various regional languages.

6.3.3 Java

Java language is used to program the application. Java is used because it is platform-independent and flexible. The most meaningful characteristic of Java is to run a program smoothly from one computer system to another.

6.3.4 XML

In developing this application, XML was used to develop the UI. The relative layout is designed so that the application can fit irrespective of the size and resolution of the mobile set.

7 Results and discussion

The mobile application is designed for smooth and easy interaction of NH people with HI people. An NH person will be able to select his or her comfortable language with the app. He or she can write a message using the search option, which is then converted to sign language and shown to the HI. Conversely, the HI person will be able to give a text input and search for the relevant sign gestures and find synonyms and antonyms along with examples of usage. Further he or she could also learn new words using this application. Additionally, synonyms, antonyms, and sentence-making examples are given so that users can easily understand the syntactic meaning of the word. A teacher can teach his or her students to increase their vocabulary with syntactic meaning, synonyms, and antonyms. A snapshot of the language selection screen is shown in Fig. 12A. The search engine in English to Bengali is shown in Fig. 12B. The app, apart from having a vast database of words provided through two different sections (General Purpose Dictionary and Academic Purpose Dictionary) also includes two other sections that are completely devoted to improving cognition in children by using word-antonym pairs. The study of general antonyms section is shown in Fig. 13A. The snapshot for mathematical opposite words can be seen in Figs. 13B and 15A and B, while general-purpose opposite words are given in Fig. 14A and B, Fig. 13C and D, the part of the app, which is essential to develop a sharp logical reasoning and a tight

(A)   (B)

Fig. 12 (A) Language selection option. (B) Search engine for English to Bengali dictionary.

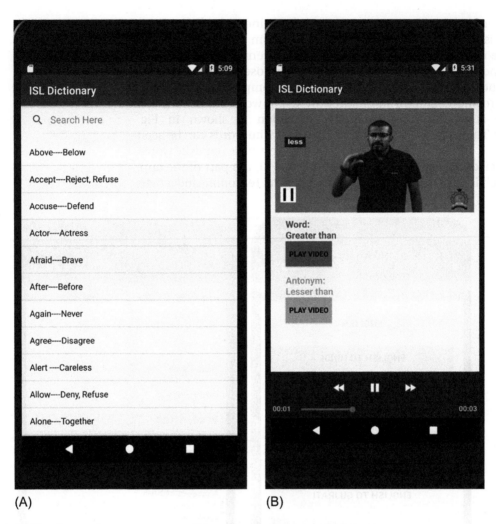

(A) (B)

Fig. 13 (A) List of general-purpose antonyms; (B) mathematical contrast words;

(Continued)

grasp on their second language (L2). This basic foundation will help such individuals to understand and expand their linguistic skills and hence allow them to overcome the mundane as well as advanced academic challenges that they presently face.

7.1 Initial verification stage

Without any errors or bugs, the app overcomes unit test as well as integration test. The app worked accurately on different screen sizes with varying resolutions on various models of the same

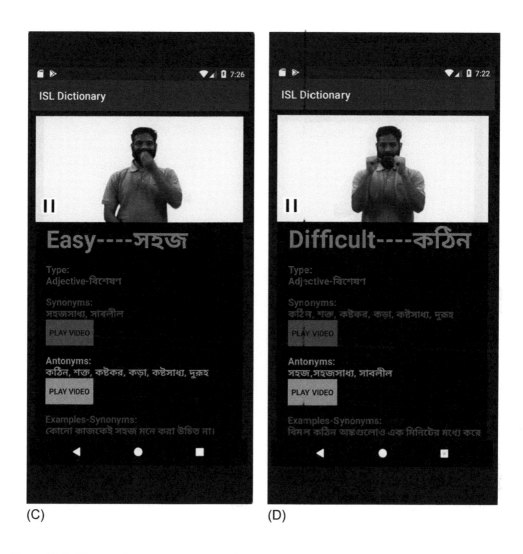

(C)   (D)

Fig. 13, cont'd (C, D) general-purpose contrast words.

device (Lenovo, Redmi, Samsung Galaxy, etc.) operating on different versions of Android (KitKat, Lollipop, Marshmallow, etc.).

7.2 Subsequent verification stage (initial acceptance testing)

The first official presentation of the app was made at a workshop on application of E-learning Software on ISL held on February 15, 2019, which was organized jointly by AYJNISHD and CDAC,

Fig. 14 (A) and (B) represent two sample opposite words with synonyms, antonyms, and uses.

Kolkata, in collaboration with the Department of Mass Education, Government of West Bengal. It was demonstrated before teachers, special educators, and students of various deaf schools all over West Bengal including some remote areas. The app was widely acknowledged and accepted by the deaf community present at the workshop.

8 Conclusion

In this chapter, we presented an efficient application to understand and develop the cognitive skill of HI students. The proposed remedial app is an outcome of the analysis of the results from a

Chapter 3 Difficulty in language comprehension and arithmetic word problems 53

Fig. 15 (A) and (B) represent two sample opposite words with synonyms, antonyms, and uses.

survey conducted with HI learners from some Bengali medium schools in and around Kolkata meant to probe into the particular necessities and difficulties faced in learning by hearing-challenged students. We focused on the concept of VKA through the understanding of antonyms on one hand, and at the same time, linking to contrasting mathematical terms to develop mathematical and logical reasoning ability, on the other hand. This application aims to help speech, and HI students by providing them with an attractive communication and learning tool. It will develop their thinking skills with logical reasoning. In this app (Fig. 16), by providing two words with opposite meaning (e.g., big-small, old-young, cry-laugh, fast-slow) simultaneously, we

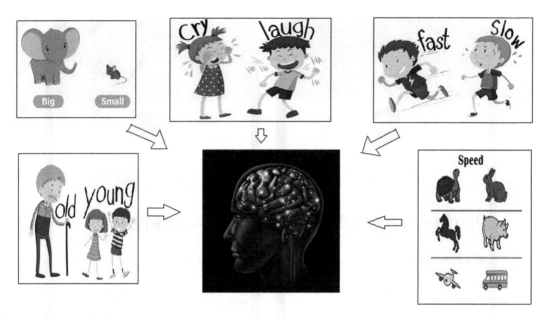

Fig. 16 Thinking skill development.

try to improve the thinking capability of the brain. The experiment intends to develop a mathematical dictionary as well as a general dictionary that emphasizes antonyms with corresponding examples for HI students so that they develop their reasoning and problem-solving capability in comprehensive arithmetic, mensuration, and various numeric problems, and build an in-depth knowledge base in the field of linguistics acquisition. New and attractive videos need to be recorded consisting of more contrasting words from mathematics (e.g., prime and composite numbers, variable and constant, etc.) along with developing a general dictionary that emphasizes antonyms with corresponding examples so that they develop their reasoning and problem-solving capability, which, in turn, further enhances their depth of language comprehension.

Acknowledgments

The work is funded by the DST Women Scientist-B scheme, Government of India, Reference No. SR/WOS-B/140/2016. The authors would also like to thank the Ramkrishna Mission Vivekanada Educational and Research Institute, Coimbatore Campus, the Indian Sign Language Research and Training Center, New Delhi, the Ali Yavar Jung National Institute of Speech and Hearing Disabilities (AYJNISHD) Eastern Regional Center, Kolkata, the Ramakrishna Vivekananda Mission,

Suryapur Center, West Bengal, the Calcutta Deaf and Dumb School, the Bali Jora Aswatthatala Vidyalaya, and many colleagues at the Indian Statistical Institute and C-DAC Kolkata for their continuous help in survey and analysis and through useful discussions.

References

[1] U. Zeshan, M.N. Vasishta, M. Sethna, Implementation of Indian Sign Language in educational settings, Asia Pac. Disabil. Rehabil. J. 16 (1) (2005) 16–40.
[2] S. Roy, A new assistive technology in Android platform to aid vocabulary knowledge acquirement in Indian Sign Language for better reading comprehension in L2 and mathematical ability, in: 2019 6th International Conference on Signal Processing and Integrated Networks (SPIN), IEEE2019, pp. 408–413.
[3] S. Roy, An app. based unified app. roach to enhance language comprehension and mathematical reasoning ability of the hearing impaired using contrast words, in: IEEE TENCON, Kochi, Kerala, India, 17th to 20th October2019.
[4] M. Marschark, Raising and Educating a Deaf Child: A Comprehensive Guide to the Choices, Controversies, and Decisions Faced by Parents and Educators, 2nd edn., Oxford University Press, Printed in United States of America, 2007.
[5] A.R. Lederberg, P.E. Spencer, Vocabulary development of deaf and hard of hearing children, Context Cogn. Deaf. (2001) 88–111.
[6] V.R. Charrow, J.D. Fletcher, English as the second language of deaf children, Dev. Psychol. 10 (4) (1974) 463–470.
[7] F.G. Birinci, The Effectiveness of Visual Materials in Teaching Vocabulary to Deaf Students of EFL (MS thesis), Eğitim Bilimleri Enstitüsü, 2014.
[8] P. Arnold, Deaf children and mathematics, Croatian Rev. Rehabil. Res. 32 (1996) 65–72.
[9] M. Hyde, R. Zevenbergen, D. Power, Deaf and hard of hearing students' performance on arithmetic word problems, Am. Ann. Deaf 148 (1) (2003) 56–64.
[10] K.L. Kritzer, C.M. Pagliaro, An intervention for early mathematical success: outcomes from the hybrid version of the building math readiness parents as partners (MRpp.) project, J. Deaf Stud. Deaf Educ. 18 (1) (2013) 30–46.
[11] Y. Zarfaty, T. Nunes, P. Bryant, The performance of young deaf children in spatial and temporal number tasks, J. Deaf Stud. Deaf Educ. 9 (3) (2004) 315–326.
[12] A. Edwards, L. Edwards, D. Langdon, The mathematical abilities of children with cochlear implants, J. Child Neuropsychol. 19 (2) (2013) 127–142.
[13] G.P. Ouellette, What's meaning got to do with it: the role of vocabulary in word reading and reading comprehension, J. Educ. Psychol. 98 (3) (2006) 554–566.
[14] M.S. Gazzaniga, G.A. Miller, The recognition of antonyms by a language-enriched right hemisphere, J. Cogn. Neurosci. 1 (2) (1989) 187–193.
[15] M. Sénéchal, G. Ouellette, D. Rodney, The misunderstood giant: on the predictive role of early vocabulary to future reading, in: Handbook of Early Literacy Research2006, pp. 173–182.
[16] P.V. Paul, J.P. O'Rourke, Multimeaning words and reading comprehension: implications for special education students, Remedial Spec. Educ. 9 (3) (1988) 42–52.
[17] M. Doherty, J. Perner, Metalinguistic awareness and theory of mind: just two words for the same thing? Cogn. Dev. 13 (3) (1998) 279–305.
[18] E.V. Clark, On the child's acquisition of antonyms in two semantic fields, J. Verbal Learn. Verbal Behav. 11 (6) (1972) 750–758.
[19] S. Jones, M.L. Murphy, Using corpora to investigate antonym acquisition, Int. J. Corpus Linguist. 10 (3) (2005) 401–422.

[20] C.I. Phillips, P. Pexman, When do children understand 'opposites'?. J. Speech Lang. Hear. Res. 58 (2015) https://doi.org/10.1044/2015-JSLHR-L-14-0222.
[21] J.Z. Sarant, C.M. Holt, R.C. Dowell, F.W. Rickards, P.J. Blamey, Spoken language development in oral preschool children with permanent childhood deafness, J. Deaf Stud. Deaf Educ. 14 (2) (2009) 205–217.
[22] Bangla Ishara Vashar Abidhan, Edited by Bangladesh Sign Language Committee, https://dokumen.tips/documents/bangla-sign-language-dictionary.html.

4

Machine learning in healthcare toward early risk prediction: A case study of liver transplantation

Parag Chatterjee[a,b], **Ofelia Noceti**[c], **Josemaría Menéndez**[c], **Solange Gerona**[c], **Melina Toribio**[b], **Leandro J. Cymberknop**[a], and **Ricardo L. Armentano**[a,b]

[a]*National Technological University (Universidad Tecnológica Nacional), Buenos Aires, Argentina.* [b]*University of the Republic (Universidad de la República), Montevideo, Uruguay.* [c]*Military Hospital (Dirección Nacional de Sanidad de las Fuerzas Armadas), Montevideo, Uruguay*

1 Introduction

Healthcare services have seen a strong paradigm shift in recent years globally. At the clinical level, personalized care to individuals is usually provided based on medical history, examination, vital signs, and evidence. However, in the recent times, the focus on these traditional tenets is being taken over by the aspects of learning, metrics, and quality improvement [1]. The last decade has seen a global rise in adoption of Electronic Health Records (EHRs) [2–7], catalyzing the increase in the complexity and volume of the health data generated in the process. Apart from the EHR-sourced ordinary patient data, due to the change in treatment paradigms and focus on lifestyle and comprehensive healthcare, new varieties of health-data about medical conditions, lifestyle, underlying genetics, medications, and treatment approaches also showed a paramount rise. Despite the complex nature of new-generation health data, human cognition to analyze and make sense of these humongous data is finite [8]. The traditional medical models of analysis deserve a reengineering for more efficiency, leading to the computer-assisted methods to organize, interpret, and recognize patterns from these data [9, 10]. Efficient collection and accurate analysis of data are critical to improvements in the effectiveness and efficiency of healthcare delivery [1]. In this respect, emerging as a promising field, eHealth addresses multifarious aspects of the healthcare system, like tracking changes in health behavior and prevention and management of chronic diseases [11].

In the recent years, the intrinsic power of data in healthcare started unveiling like never before, leading to the endeavor in making sense of health data in the best possible way, using advanced data analytics and computational intelligence. Especially in the field of healthcare, the aspect of intelligent data analytics is one of the most trending topics worldwide [12]. One of the prime areas where such analyses have been applied is the field of chronic diseases. By 2020, chronic diseases are expected to contribute to 73% of all deaths worldwide and 60% of the global burden of disease. Also, 79% of the deaths attributed to these diseases occur in the developing countries. Four of the most prominent chronic diseases—cardiovascular diseases (CVD), cancer, chronic obstructive pulmonary disease, and type 2 diabetes—are linked by common and preventable biological risk factors, notably high blood pressure, high blood cholesterol, and overweight, and by related major behavioral risk factors. To prevent these major chronic diseases, the actions need to be centered around controlling the key risk factors in a comprehensive and integrated manner [13]. In addition to the chronic diseases, the aspect of intelligent risk prediction and preventive actions count significant even in the domain of transplantations [10, 14].

Moreover due to the concept of context awareness backed by sensor fusion in the environment of smart eHealth systems and IoT, the health data generated and acquired is more comprehensive and detailed. The smart prediction and prevention systems in healthcare usually share some common steps like the collection of health data from sensors or other sources and assimilating the EHRs, followed by analyzing and computing the risks [15]. In the endeavor of taking possible actions to prevent chronic diseases, detecting the diseases at an early stage stands to be the prime challenge. Most of these diseases do not exhibit clearly identifiable signs at the early stage. And here lies the key area of Artificial Intelligence (AI), harvesting the possibility of early detection of these diseases in terms of risk. From the data science perspective, the aspect of health data acts as the key valuable resource. Most importantly, the domain of health data has expanded dramatically over these years. Even the superficially and noninvasively obtained behavioral, physiological, and metabolic health data hold enormous significance. In the domain of early detection and prediction of diseases, the health data possesses a huge potential. Disease prediction led by AI is a multilevel process. It involves the analysis of the intricate details inside the health data, looking for early indications or traces of diseases.

Thanks to the recent boost in paradigms like IoT, eHealth, and medical informatics pertaining to AI, healthcare is one of the most important areas where data analytics finds its applications and analyzing health data has reached new heights [15a]. Minimizing the response time in diagnosis and treatment is a crucial component in efficient healthcare services, which makes the power of data analytics relevant for faster analysis and use of intelligent methods for better diagnosis [16]. Detailed analysis of health data expedites the automated diagnosis on one hand, also leading to personalized treatment. On the other hand, it provides the comprehensive and holistic information of a large group of people under treatment. This is fairly advantageous to automate the process of monitoring along with prediction of health risks obtained from the analysis of the health data of the patients. In this direction, one of the key areas where prediction of risk is highly crucial is transplantation. Transplantation is itself a complex procedure that makes the consideration of pretransplant predictions of risk an important factor, opening up a broad scope for computational intelligence. Possible insights obtained from the pretransplant health data of the patients by harnessing the power of AI would positively influence the healthcare delivery approach. In this work a case study is presented highlighting the aspects of computational intelligence toward risk prediction in liver transplantation.

Liver transplantation is the last therapeutic option in patients with end-stage liver disease. Being a complex healthcare process, it is related to humongous costs and requires the expertise of a specialized interdisciplinary team along with a close monitoring of patients during the entire timeline. This process generates a large volume of complex, multidimensional data. The adequate clinical management of transplant patients impacts their vital prognosis, and decisions on many occasions are made from the interaction of multiple variables [17]. However, there exists an enormous demand in the domain of prediction in liver transplantation process, ranging from survival till the suitability of transplant [18].

The healthcare sector has emerged as one of the prime areas to adopt new technologies, given that the primary objective is to provide better and more efficient treatment to the patients. Healthcare delivery is a complex aspect at both individual and population levels. From the perspective of data-driven insights for any medical personnel, a large pool of historical health data of the patient is a huge plus, before starting a thorough treatment [19].

At the clinical level the aspect of providing healthcare services is guided mostly by medical history, examination, vital signs, and evidence. Recent times have seen a paradigm-shift of the core traditional approaches toward the supplementary focus on learning, metrics, and quality improvement of the healthcare provided. The collection and analysis of good-quality data are critical to improvements in the effectiveness and efficiency of delivering healthcare services [20]. The field of artificial intelligence applies to a wide range of disciplines in medicine; however, in transplantation, it is still a scarcely explored area.

The AI has started playing an important role in predicting the main determinants of morbidity and mortality in patients, which stands quite significant in the domain of transplantations. This deals with analyzing the probability of developing an inherent risk of disease or complication during the entire timeline. The main objective of this work is to spotlight the applied aspects of data analytics in healthcare and importance of AI in transplantation, illustrated through a case study of liver transplantation at the National Center for Liver Transplantation and Liver Diseases, Uruguay. Also, based on the advantages of AI in transplantation, an AI-based predictive clinical decision support system for transplantations has been proposed for early detection and prediction of risks and proffering better diagnosis and treatment to the patients.

2 Background of the study: Description of cohort

This study is focused on the patients registered under the National Liver Transplantation Program in Montevideo, Uruguay. In this case the patients considered for the study were registered into the program between the years 2014 and 2017 and were evaluated at the time of their registration to the program. Depending on the assessment at the time of registration and assessing the severity of their illness and comorbidities, the patients qualify to continue in the registered waitlist for liver transplantation or step down from the list. Based on further analysis, some patients from the pool of enlisted patients proceed to liver transplantation, whereas some drop out from the list due to progression illness and the rest continue in the waitlist. To arrange and prioritize the patients in the waitlist, the Model for End-Stage Liver Disease (MELD) score is used, in accordance to most of the liver-treatment centers. 104 patients consist of the cohort considered here (Table 1). The mean age of the population was 47 years (ranging from 14 to 70 years), with almost

Table 1 Details of health parameters of the cohort.

Parameters	Cohort properties
Total number of patients	104
Age at the moment of evaluation (years)	47 ± 15
Gender (%)	Male: 51% | Female: 49%
BMI (kg/m^2)	27 ± 5
Systolic blood pressure (mm Hg)	117 ± 12
Diastolic blood pressure (mm Hg)	67 ± 8
Total Cholesterol (mmol/L)	159 ± 93
Triglycerides	106 ± 71
HDL (mmol/L)	38 ± 24
LDL (mmol/L)	97 ± 60
Total cholesterol/HDL	12 ± 26
Platelets (×1000)	125 ± 77
Lymphocytes	1333 ± 854
Neutrophils	3722 ± 2040
Monocytes	599 ± 343
Eosinophils	203 ± 249
Basophils	21 ± 45
Glycemia	100 ± 40
Smoking	Yes: 25% | No: 75%
Diabetes	Yes: 23% | No: 77%
Hypertension	Yes: 28% | No: 72%

Values are expressed as Mean ± Standard Deviation.

equal number of males and females. The most frequent indication for liver transplantation was cirrhosis, followed by hepatocellular carcinoma and acute liver failure.

The aspect of risks constitutes an entire domain. Some the usual risk scores in this case were dependent on the health parameters. For example, Framingham Cardiovascular Disease 10-year risk (FR) was taken into account as one of the interesting risk factors. It mostly considered the parameters of age, gender, blood pressure, LDL and HDL cholesterol, smoking, and diabetes. Another risk score considered was the MELD, which took into account dialysis information and parameters like creatinine, bilirubin, INR, and sodium. Apart from that, parameters like death (dead/alive) and transplant (transplanted/waitlisted) were also considered interesting for analysis as dependent variables for this study [17].

3 Intelligent risk analysis: Aspect of vascular age and cardiometabolic health

The original focus of this study was to analyze all the health parameters of the cohort and obtain interesting relationships and correlations among the health parameters and the risk scores, especially with the intention to analyze the early detection and prediction of risks in a pretransplant scenario.

One of the most crucial tasks in this respect was to identify the most suitable data model from measurements of the system inputs and outputs. Especially in the field of diseases and risk prediction, the data handled are mostly multidimensional [10]. Similarly, in this case, the dataset constituted by the cohort from the transplantation program had more than 20 dimensions. Several methods were analyzed in the beginning, before actually choosing one. In the initial phase of data preprocessing and data organization, principal component analysis (PCA) was considered to be used as a dimensionality reduction technique. But though the original intention was to take advantage of the main benefit to PCA, which is to reduce the size of the feature vectors for computational efficiency, it results in loss of important information, especially taking into account several health parameters, have quite small impact on the risk scores, but still stands interesting from a medical point of view. Also, from the perspective of computational intelligence, several supervised learning methods were not considered viable in this specific case because of their complexity and training times and their background of supervised learning [21]. In this respect, unsupervised learning techniques were of key interest, since the intention was to analyze different groups among the cohort and decipher interesting insights. A considerable part of data in such cases often arriving unlabeled, unsupervised learning methods help in finding patterns in the data or to analyze the health scenario over a big population. In this aspect, clustering techniques like k-means often stand very useful in separating a group of patients into different clusters and then to analyze in detail the salient features and distinct characteristics [10]. Briefly speaking, k-means clustering aims to find the set of k clusters such that every data point is assigned to the closest center and the sum of the distances of all such assignments is minimized [22]. Especially when the relationships and impact of different health parameters are not known well, clustering techniques are often used to separate a patient population to study the influencing factors. In this aspect, k-means clustering counts useful because of its simplicity and was chosen to be applied in this study of the cohort.

In the initial phase the key objective was to analyze the impact and relevance of each health parameter with respect to the risk scores. Despite that some of the parameters showed good relationships with the risk scores, some parameters were itself included in the calculation of the risk scores; of course, tautology was evident. Also the focal goal was to understand the overall risk scenario of the cohort, considering all the health parameters and relating it with the risk scores. Aiming at analyzing the full patient health profile, clustering was performed on the entire cohort taking all the features into account. To fix the missing values in some features, data cleaning was performed. However, in this case, the risk parameters like FR, MELD, and dependent parameters like death and transplantation events were also included in this clustering. The silhouette criterion indicated that the optimum number of clusters is two (Fig. 1). After performing the clustering with two clusters, each of the clusters illustrated distinguishing characteristics between each other (Table 2).

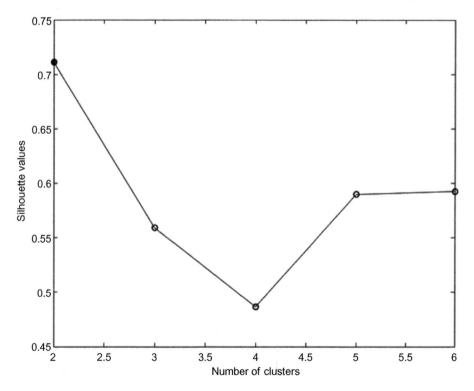

Fig. 1 Silhouette criterion (optimum number of clusters = 2).

Table 2 Cluster characteristics (all patients, all parameters).

Parameters	Cluster 1	Cluster 2
Total number of patients	72	25
Age at the moment of evaluation (years)	46 ± 16	46 ± 16
Gender (%)	Male: 54%	Male: 48%
	Female: 46%	Female: 52%
BMI (kg/m^2)	27 ± 5	26 ± 5
Systolic blood pressure (mm Hg)	117 ± 11	116 ± 14
Diastolic blood pressure (mm Hg)	67 ± 8	68 ± 8
Total cholesterol (mmol/L)	150 ± 54	184 ± 162
Triglycerides	92 ± 42	149 ± 115
HDL (mmol/L)	42 ± 24	29 ± 24
LDL (mmol/L)	90 ± 44	113 ± 92
Total cholesterol/HDL	7 ± 12	26 ± 47
Platelets (x1000)	107 ± 58	174 ± 97
Lymphocytes	1225 ± 799	1568 ± 908
Neutrophils	2794 ± 1067	6460 ± 1688
Monocytes	524 ± 287	808 ± 413
Eosinophils	194 ± 191	240 ± 382
Basophils	18 ± 42	28 ± 54
Glycemia	98 ± 39	103 ± 45
Framingham risk	6 ± 5	8 ± 7
MELD	16 ± 6	20 ± 9
Death	Dead: 14%	Dead: 32%
Transplantation	Transplanted: 58%	Transplanted: 68%

Values are expressed as Mean ± Standard Deviation/Ratio/Percentage.

In the cohort considered in this study, there are patients in the registered list who eventually had liver transplants and patients who continued in the waitlist. Also among each group, the patients were separated into alive and dead to study their respective physiological characteristics. The motive was to assess the trends in physiological parameters of the groups and also to study their distinguishable risks.

Despite the clusters showing quite interesting characteristics, the process itself included several risk scores, or rather dependent variables (e.g., MELD, death, transplantation, and FR). Intuitively, it was evident that these scores also had a significant impact in affecting the separation of the cohort into two clusters. To reduce the impact of the risk scores itself in the clustering process and to

Table 3 Cluster characteristics (all patients, without "parameters of interest").

Parameters	Cluster 1	Cluster 2
Total number of patients	61	36
Age at the moment of evaluation (years)	46 ± 16	48 ± 14
Gender (%)	Male: 49%	Male: 55%
	Female: 51%	Female: 45%
BMI (kg/m^2)	27 ± 5	27 ± 5
Systolic blood pressure (mm Hg)	117 ± 12	116 ± 12
Diastolic blood pressure (mm Hg)	67 ± 8	68 ± 8
Total Cholesterol (mmol/L)	148 ± 58	175 ± 136
Triglycerides	91 ± 40	134 ± 102
HDL (mmol/L)	42 ± 24	31 ± 24
LDL (mmol/L)	88 ± 45	109 ± 79
Total cholesterol/HDL	9 ± 18	20 ± 40
Platelets (x1000)	102 ± 59	165 ± 86
Lymphocytes	1100 ± 679	1667 ± 961
Neutrophils	2616 ± 1237	5808 ± 1735
Monocytes	475 ± 274	825 ± 383
Eosinophils	187 ± 202	233 ± 323
Basophils	11 ± 32	36 ± 59
Glycemia	100 ± 40	97 ± 42

analyze the clusters from an independent and less biased point of view, four parameters (FR, MELD, death, and transplantation) were not included in the clustering process. This clustering also separated the cohort into two clusters with the following properties (Table 3). The silhouette plot using the Euclidean distance metric showed that the data are split into two clusters, one smaller and one bigger, similar to the clusters obtained with all patients and all parameters. Most of the points in the two clusters had large silhouette values), indicating that the clusters were separated quite well. However, since the four parameters were not considered in the clustering process, after the clusters were obtained, those four parameters were linked to the corresponding patients, and their respective characteristics were analyzed (Table 4) [17].

A similar approach was followed to perform clustering separately on the waitlisted patients and the transplanted patients,

Table 4 Postclustering analysis (all patients—FR, MELD, death, and transplantation).

Parameters	Cluster 1	Cluster 2
Framingham risk	6 ± 5	7 ± 6
MELD	16 ± 6	19 ± 8
Death	Dead: 16%	Dead: 25%
Transplantation	Transplanted: 54%	Transplanted: 69%

but without including FR, MELD, death, and transplantation condition in the clustering process. However, in the same manner, those were analyzed after the clustering process with the corresponding patients. In each case of waitlisted and transplanted patients, two clusters were obtained with distinguishable characteristics. Silhouette analysis was performed in each case (transplant patients and waitlist patients, all parameters except the four parameters of interest), and the cluster points indicated that the clusters were separated quite well. Also the four risk parameters of interest were analyzed in relation to the first two cluster-analyses performed on all the patients [17].

In an extended version of this study, the aspect of risks has been widened to the domain of cardiovascular health. Human body being a comprehensive and complex structure, the important relationship between the liver and the circulatory system is quite evident. The latter, made up of the heart and all the body's blood vessels, is affected by various parameters that can impair or improve its functioning and, therefore, the health of the individual. This leads to focusing on the state of the patients' cardiovascular health, to have a snapshot of the relationship between the parameters that interfere with it and the health risks of liver patients as well. In this respect the Framingham heart study was used as the model to analyze the cardiovascular risks of the cohort. The 10-year cardiovascular disease risk function [23] was used, considering the parameters like age, gender, total cholesterol, HDL, hypertension, diabetes, and smoking, which in turn returned the Framingham Risk score, along with the vascular age of the respective patient. The percentage of risk (10-year prediction) considered here reflects the possibility of suffering a cardiovascular event like coronary death, myocardial infarction, coronary insufficiency, angina, ischemic stroke, hemorrhagic stroke, transient ischemic attack, peripheral arterial disease, or heart failure. Specifically in this case an extended cohort of 170 patients evaluated between 2014 and 2019 were considered, where 55% of patients were transplanted, and the remaining 45% were divided into waitlisted patients, who were removed from the list due to improved health, or due to death after the evaluation and before receiving the transplant. In this cohort the risk function was applied, and Framingham risk score and vascular age for the patients were calculated. Also, for each patient, the difference between their vascular age and biological age was calculated. Based on the ΔAge (Vascular Age – Biological Age), the cohort was separated into two groups, where ΔAge > 0 (Vascular Age $>$ Biological Age) and ΔAge ≤ 0 (Vascular Age $\leq$ Biological Age).

4 Results and discussions

The general observation from the clusters obtained after the entire cluster analysis performed on various instances, every time two clusters were obtained, containing always a bigger cluster and a smaller cluster.

In the first study, where clustering was performed on the all patients and all parameters, the smaller cluster showed higher mean value of FR (8) and MELD (20) than the bigger cluster (FR: 6, MELD: 16). Also the smaller cluster showed higher percentage of death (32%) and higher percentage of transplant patients (68%) than the bigger cluster (death: 14%, transplantation: 58%). Apart from that, though both the clusters have almost similar mean age, BMI, and blood pressure, the smaller cluster showed significantly lower value of HDL and higher values for LDL, triglycerides, total cholesterol/HDL, platelets, lymphocytes, neutrophils, monocytes, eosinophils, basophils, and glycemia, implying the smaller cluster at higher risk than its bigger counterpart.

In the second case, when the clustering was performed on all the patients but without considering the FR, MELD, death, and transplantation, two clusters were obtained, one bigger (61 patients) and the other smaller (36 patients). Here also the smaller cluster showed higher values of FR (7), MELD (19), and higher percentage of death (25%) and transplant patients (69%) than the bigger cluster (FR: 6, MELD: 16, death: 16%, transplantation: 54%). Also, though both the clusters have almost similar mean age, BMI, and blood pressure, the smaller cluster showed significantly lower value of HDL and higher values for LDL, triglycerides, total cholesterol/HDL, platelets, lymphocytes, neutrophils, monocytes, eosinophils, basophils, and glycemia, implying the smaller cluster at higher risk than the bigger one.

The similar trend followed in the clusters obtained from the waitlisted patients and the transplant patients. In all the cases, apart from the risk scores (FR and MELD) and parameters like death and transplantations, other parameters like lymphocytes, neutrophils, and monocytes showed a significant ($P < 0.05$) rise in the smaller cluster than the bigger one. From the medical perspective, higher lymphocyte count indicates to possibilities of lymphocytosis, frequently associated with chronic infections, inflammations, and autoimmune diseases. Also, higher count of monocytes indicates to potential risk of infection and neutrophil and platelet count signals to the inflammation status.

Intuitively, it turned out that after every clustering, among the two clusters obtained, the smaller cluster demonstrates more risk and more vulnerable patients than the ones in the bigger cluster. Also it implied that at the point of evaluation in the timeline, with no knowledge of the future events, the patient population could be divided successfully into two clusters, considering just the parameters obtained during evaluation. Even without taking into account the risk parameters like FR and MELD, the patient population could be separated into two groups, with one of those showing significantly higher risks than the other. For example, the mean time to transplant from the point of evaluation being 3 months (the maximum being 3 years), this clustering model could direct a patient to a cluster with high risk or low risk, just using the parameters at the time of evaluation, while enlisting into the system. Such approach is apt for inclusion in clinical decision support systems in transplantations. A typical non–knowledge based clinical decision support system (CDSS) holds in its core the power of machine learning. Feeding the comprehensive health records of the patient cohort into the CDSS makes it possible for artificial intelligence to analyze the cohort and classify them into risk groups, eventually alerting the medical personnel the dynamic possibilities of complications for a particular patient in a pretransplant scenario. Thus, it shows the holistic view to the overall health conditions of a group of people, facilitating the identification of high and risk groups [17, 24].

In the extended study of the cardiometabolic risk groups within the patient cohort, the separated groups based on their ΔAge (Vascular Age − Biological Age) showed distinguishing properties with respect to their health parameters (Table 5).

Analyzing the two groups with respect to their ΔAge (Vascular Age − Biological Age), it is quite natural that the group of patients with ΔAge > 0 has higher BMI and higher percentage of smoking, diabetes, and hypertension, along with significantly high values of total cholesterol/HDL than the group of patients with ΔAge ≤ 0. This can be explained in terms of cardiometabolic risks due to the presence of those parameters within the calculation of vascular age itself. But in addition to that, the first group with ΔAge > 0 showed higher number of MELD score and higher percentage of transplantations. The MELD score estimates a patient's chances

Table 5 Risk-group properties (separated cohort in terms of vascular age).

Parameters	Vascular age > biological age	Vascular age ≤ biological age
Total number of patients	75% of the cohort	25% of the cohort
Age at the time of evaluation (years)	48 ± 15	49 ± 12
Gender	Male: 59% | Female: 46%	Male: 39% | Female: 61%
BMI (kg/m^2)	28 ± 6	26 ± 4
Systolic blood pressure (mm Hg)	119 ± 14	106 ± 13
Diastolic blood pressure (mm Hg)	69 ± 10	63 ± 8
Total cholesterol (mmol/L)	159 ± 93	143 ± 49
Triglycerides	117 ± 82	96 ± 67
HDL (mmol/L)	30 ± 21	47 ± 26
LDL (mmol/L)	103 ± 60	78 ± 32
Total cholesterol/HDL	14 ± 27	4 ± 4.4
Platelets (×1000)	119 ± 76	110 ± 54
Lymphocytes	1435 ± 1476	1356 ± 978
Neutrophils	4199 ± 3691	3924 ± 3549
Monocytes	623 ± 453	514 ± 321
Eosinophils	189 ± 239	150 ± 158
Basophils	22 ± 46	8 ± 28
Glycemia	108 ± 48	91 ± 21
Smoking	Yes: 32% | No: 68%	Yes: 3% | No: 97%
Diabetes	Yes: 39% | No: 61%	Yes: 19% | No: 81%
Hypertension	Yes: 34% | No: 66%	Yes: 14% | No: 86%
MELD Score	18 ± 7	16 ± 8
Framingham CV risk score	15 ± 13	5 ± 3
Vascular age	62 ± 19	43 ± 12
Vascular Age–Biological Age	14 ± 9	−7 ± 4
Transplantation	Yes: 62% No: 38%	Yes: 56% No: 44%

of surviving their disease during the next 3 months, and physicians working on the liver transplant program in this case used MELD score to classify the level of liver severity of patients and determine the urgency of the liver transplant [25], with a higher MELD indicating higher severity. It has been observed in this study that the group of patients having their vascular age more than their actual biological age also showed higher severity with respect to MELD score and eventually showed higher percentage of transplantations, indicating a possible relationship of transplantations with cardiometabolic risks.

The study and the analysis invoke the possibility of evaluation of a new patient entering the evaluation list for the liver transplantation program using a CDSS, to predict the cardiometabolic risks along with the usual evaluation procedure and instantaneously assign the risk group the new patient falls in. This would be a key aspect of a potentially assistive tool to the medical personnel to classify the patient cohort into risk groups at any given time starting from the entry-point evaluation.

5 Conclusion

In the new age of data and eHealth, the inherent knowledge of data has turned out to be of immense importance, and computational intelligence plays a key role in making the most out of the data. Especially for chronic diseases, long-term behavioral and lifestyle data stand quite crucial. Data modeling and predictive analytics open a huge avenue toward clinical decision support systems, which is a fundamental tool nowadays for preventive and personalized healthcare and supports healthcare providers to have deeper insights into patients' data [26] and take clinical decisions [27]. Transplantation are associated with several risk factors, which if predicted better using computational intelligence, stands quite paramount in reducing the mortality in transplantation process. In this case study of liver transplantation, it was important to ascertain the suitability to perform the transplantation, and this invokes the need of analysis of other possible risks of the patient before taking the decision. Despite the presence of definite risk scores to determine the severity of the health conditions of a patient in the list for transplant, computational intelligence leaves scope to take advantage of all the health parameters evaluated in ascertaining the risks with higher efficiency. This work takes a step in that direction, using the simple aspects of artificial intelligence in segregating the patient cohort into risk groups from a predictive point of view, considering the simple health parameters, which are normally evaluated for every enlisted patient. Though based on the volume of the cohort in this work, it is difficult to design precise risk-model, but the inferences could be scaled to a larger population leaving the scope to validate in a larger cohort as well. Decision support systems are interesting components of recent healthcare systems, which analyze data and support healthcare providers to take clinical decisions [26, 27]. This work leads to the idea of a predictive clinical decision support system, aiming at automatically classifying the patient population at the evaluation time into high risk or low risk, facilitating the aspect of care during the enlisted period [17].

Acknowledgments

This work is financially supported by the National Agency for Research and Innovation (*Agencia Nacional de Investigación e Innovación*) of Uruguay through its grant FSDA-2017-1-143653 (*Inteligencia Computacional en Salud. Creando Herramientas para Predecir la Sobrevida en Trasplante Hepático*).

References

[1] WHO. Bull. World Health Organ. 2015;93:203–208. https://doi.org/10.2471/BLT.14.139022 [Internet]. 2019. Available from: https://www.who.int/bulletin/volumes/93/3/14-139022/en/.

[2] Charles DK J, Patel V, Furukawa M. Adoption of Electronic Health Record Systems among U.S. Non-federal Acute Care Hospitals: 2008–2012; 2013. http://www.healthit.gov/sites/default/files/oncdatabrief9final.pdf. Accessibility verified April 20, 2014.

[3] N.H. Shah, Translational bioinformatics embraces big data, Yearb. Med. Inform. 7 (1) (2012) 130–134.

[4] O. Heinze, M. Birkle, L. Koster, B. Bergh, Architecture of a consent management suite and integration into IHE-based regional health information networks, BMC Med. Inform. Decis. Mak. 11 (2011) 58.

[5] A. Tejero, I. de la Torre, Advances and current state of the security and privacy in electronic health records: survey from a social perspective, J. Med. Syst. 36 (5) (2012) 3019–3027.

[6] A. Mense, F. Hoheiser-Pfortner, M. Schmid, H. Wahl, Concepts for a standard based cross-organisational information security management system in the context of a nationwide EHR, Stud. Health Technol. Inform. 192 (2013) 548–552.

[7] A. Faxvaag, T.S. Johansen, V. Heimly, L. Melby, A. Grimsmo, Healthcare professionals' experiences with EHR-system access control mechanisms, Stud. Health Technol. Inform. 169 (2011) 601–605.

[8] M.K. Ross, W. Wei, L. Ohno-Machado, "Big data" and the electronic health record, Yearb. Med. Inform. 9 (1) (2014) 97–104, https://doi.org/10.15265/IY-2014-0003.

[9] K.B. Wagholikar, V. Sundararajan, A.W. Deshpande, Modeling paradigms for medical diagnostic decision support: a survey and future directions, J. Med. Syst. 36 (5) (2012) 3029–3049.

[10] P. Chatterjee, L.J. Cymberknop, R.L. Armentano, Nonlinear systems in healthcare—a perspective of intelligent disease prediction, in: W. Legnani (Ed.), Nonlinear Systems, vol. 2, IntechOpen, 2019https://doi.org/10.5772/intechopen.88163https://www.intechopen.com/online-first/nonlinear-systems-in-healthcare-towards-intelligent-disease-prediction.

[11] D.K. Ahern, J.M. Kreslake, J.M. Phalen, What is eHealth (6): perspectives on the evolution of eHealth research. J. Med. Internet Res. 8 (1) (2006) e4, https://doi.org/10.2196/jmir.8.1.e4.

[12] V.E. Balas, S.S. Roy, D. Sharma, P. Samui (Eds.), Handbook of Deep Learning Applications, In: vol. 136, Springer, 2019.

[13] WHO. Integrated Chronic Disease Prevention and Control [Internet]. 2019. Available from: https://www.who.int/chp/about/integrated_cd/en/.

[14] L.B. VanWagner, H. Ning, M. Whitsett, J. Levitsky, S. Uttal, J.T. Wilkins, M.M. Abecassis, D.P. Ladner, A.I. Skaro, D.M. Lloyd-Jones, A point-based prediction model for cardiovascular risk in orthotopic liver transplantation: the CAR-OLT score. Hepatology 66 (2017) 1968–1979, https://doi.org/10.1002/hep.29329.

[15] Masoud Hemmatpour, Renato Ferrero, Filippo Gandino, Bartolomeo Montrucchio, and Maurizio Rebaudengo. Nonlinear predictive threshold model for real-time abnormal gait detection. J. Healthc. Eng., vol. 2018, Article ID 4750104. 9 pages. https://doi.org/10.1155/2018/4750104.

[15a] P. Chatterjee, R. Armentano, L. Palombi, L. Kun, Editorial preface: special issue on IoT for eHealth, elderly and aging. Internet of Things (2019) 100115, https://doi.org/10.1016/j.iot.2019.100115.

[16] L.E. Romero, P. Chatterjee, R.L. Armentano, An IoT approach for integration of computational intelligence and wearable sensors for Parkinson's disease diagnosis and monitoring, Health Technol. 6 (3) (2016) 167–172, doi: 10.1007/s12553-016-0148-0.

[17] P. Chatterjee, et al., Predictive risk analysis for liver transplant patients—eHealth model under National Liver Transplant Program, Uruguay. IEEE 9th International Conference on Advanced Computing (IACC), Tiruchirappalli, India 2019 (2019) 75–80, https://doi.org/10.1109/IACC48062.2019.8971514.

[18] D. Bertsimas, J. Kung, N. Trichakis, Y. Wang, R. Hirose, P.A. Vagefi, Development and validation of an optimized prediction of mortality for candidates awaiting liver transplantation, Am. J. Transplant. 19 (4) (2019) 1109–1118.

[19] P. Chatterjee, L. Cymberknop, R. Armentano, IoT-Based Decision Support System Towards Cardiovascular Diseases. Córdoba, Argentina, SABI, 2017.

[20] WHO. Bull. World Health Organ. 2015;93:203–208. doi: https://doi.org/10.2471/BLT.14.139022 [Internet]. 2019. Available from: https://www.who.int/bulletin/volumes/93/3/14-139022/en/.

[21] T. Hippolyte, M. Adamou, N. Blaise, C. Pierre, M. Olivier, Linear vs non-linear learning methods—a comparative study for forest above ground biomass, estimation from texture analysis of satellite images, ARIMA J. 18 (2014) 114–131.

[22] Healthcare.ai. Step by Step to K-Means Clustering. Data Science Blog, 21 July 2017 healthcare.ai/step-step-k-means-clustering/.

[23] R.B. D'Agostino, R.S. Vasan, M.J. Pencina, P.A. Wolf, M. Cobain, J.M. Massaro, W.B. Kannel, General cardiovascular risk profile for use in primary care. Circulation 117 (6) (2008) 743–753, https://doi.org/10.1161/circulationaha.107.699579.

[24] P. Chatterjee, L. Cymberknop, R. Armentano, IoT-based ehealth toward decision support system for CBRNE events. in: A. Malizia, M. D'Arienzo (Eds.), Enhancing CBRNE Safety & Security: Proceedings of the SICC 2017 Conference, Springer, Cham, 2018. https://doi.org/10.1007/978-3-319-91791-7_21.

[25] Understanding MELD Score for Liver Transplant: UPMC. (n.d.). Retrieved March 21, 2020, from https://www.upmc.com/services/transplant/liver/process/waiting-list/meld-score.

[26] P. Chatterjee, L.J. Cymberknop, R.L. Armentano, IoT-based decision support system for intelligent healthcare—applied to cardiovascular diseases, in: 7th International Conference on Communication Systems and Network Technologies (CSNT); Nagpur, 2017, pp. 362–366, https://doi.org/10.1109/CSNT.2017.8418567.

[27] P. Chatterjee, R.L. Armentano, L.J. Cymberknop, Internet of things and decision support system for eHealth-applied to cardiometabolic diseases, in: International Conference on Machine Learning and Data Science (MLDS); Noida. Piscataway: IEEE, 2017, pp. 75–79, https://doi.org/10.1109/MLDS.2017.22.

Utilizing BERT for biomedical and clinical text mining

Runjie Zhu[a], Xinhui Tu[b], and Jimmy Xiangji Huang[c]

[a]Information Retrieval and Knowledge Management Research Lab, Lassonde School of Engineering, York University, Toronto, ON, Canada. [b]School of Computer Science, Central China Normal University, Wuhan, China. [c]School of Information Technology, York University, Toronto, ON, Canada

1 Introduction and motivation

The lack of training data has always been one of the biggest challenges to tackle and overcome in the natural language processing research community. Many of the datasets with specific tasks rely on a very limited number of human-labeled training samples. However, natural language processing is a research field with extensive distinct and diversified tasks that these human-labeled training samples are neither enough in quantity nor in quality. As the models with deep learning approaches become more prominent in the natural language processing research field, it is necessary to fill in the gap of this large amount of data needed for even higher-quality research.

For that purpose, researchers in NLP community have been working together diligently to propose novel language representation models that served for general purpose. These proposed models do the pretraining work first by making use of the large amount of unannotated text data available online, following by task-specific fine-tuning step to improve the state-of-the-art experimental results significantly.

At the end of 2018, the BERT model was introduced by Google and has made incredibly great achievements and performances in many natural language processing tasks in general language domain since then. The various NLP tasks include but are not limited to name entity recognition, relation extraction, and question answering. According to the paper, BERT has revolutionarily broke the record in 11 NLP tasks with incredibly significant absolute improvement percentages. The model has made advancements of 7.7% on GLUE, 4.6% on MultiNLI, 1.5% on SQuAD v1.1 QA Test F1, and 5.1% on SQuAD v2.0 Test F1. Moreover the BERT model is considered to be the third language representation model pretrained on plain texts, namely, Wikipedia and BooksCorpus, after ELMO and GPT, to perform deeply bidirectional transformers in an unsupervised manner.

In fact the pretrained language representations are capable of dealing with a wide range of tasks in the research field and specific research domains, including biomedical and clinical studies. These tasks include but are not limited to pretrained word embeddings, sentence embeddings, and contextual representations [1–5].

As the rapid growth of the information is available online or in almost all industries, information explosion is happening everywhere. As one of the dominant industries in this age, information in medical domain has also grown significantly fast in the past decades, especially the ones in form of text. Biomedical and clinical texts by nature are different from the texts that serve for general purposes. They are usually large in data volume, wide in data variability, low in data quality, and high in data uncertainty. Problems such as heterogeneousness, noisiness, incompleteness, and ambiguousness are so much more common compared with other domains.

Therefore, to get the best empirical experimental results in medical domain, the language representation models, or BERT-based models, constructed on general purpose text corpora are not enough. To effectively process these very specific vocabularies and grammar contained in the biomedical and clinical texts, we need special BERT models that are pretrained on biomedical and clinical contexts, instead of the original BERT, which is pretrained using Wikipedia and BooksCorpus. Given that context the biomedical and clinical NLP community researchers have been actively proposing BERT-based models to process biomedical and clinical text information more effectively and efficiently. Successful biomedical and clinical BERT models include but are not limited to SciBERT [6], which is built on basic BERT to address the performance on scientific data, released by Allen Institute for Artificial Intelligence (AI2); BioBERT [5], which is a domain-specific BERT model pretrained on large amount of biomedical text corpora; ClinicalBERT [7], which is a language representation model to extract high-quality relationships between medical concepts from extensive clinical notes; and Clinical BERT Embeddings [8], which aims to build clinical specific BERT resources publicly available to benefit the entire community.

In this chapter, we classify these BERT-based models in the medical domain into two groups, namely, pretrained BERT models and fine-tuned BERT models. In Section 2, we briefly introduce the basic BERT model and the background of the model and present some key successes BERT has achieved in general purpose language domain. In Section 3, we empirically compare the major contributions, architectures, datasets applied, and experiments conducted of these medical BERT-based models and systematically discuss the strengths and limitations of these models. In Section 4, we conclude our studies and findings and present the possible directions to deepen the biomedical and clinical text mining research with BERT in future.

2 BERT

The BERT [9] proposed by Devlin et al. is the origination of the entire BERT wave. It is a novel and powerful language representation model that is naturally bidirectional and generalizable. Furthermore, it is also designed to pretrain representations from unlabeled text corpus. The major contributions of this paper are as follows: (1) it proved the effectiveness of masked deep bidirectional language representation model, and (2) it suggests the possibility of keeping the state-of-the-art performance on specific tasks while reducing the extra heavy work on architecture adjustments.

In terms of the architecture, the BERT model uses a multilayered bidirectional transformer encoder introduced by Vaswani et al. [10] earlier in 2017. And the model is differentiated in two sizes, where L stands for number of layers, H stands for the hidden size, and A stands for the number of self-attention heads: (L, 12; H, 768; A, 12, and total parameters, 110 M) and (L, 24; H, 1024; A, 16; and total parameters, 340 M). The uniqueness is that the model can jointly consider the context wrapped in both directions, left and right, in all layers to produce the deep representations. To fulfill certain specific natural language processing (NLP) tasks, the model can simply achieve it by adding one single output layer on top of the basic architecture without substantial and complex work on model redesigns or modifications. Specifically the model simply takes a sentence as an input and produces a group of vectors on each word of the sentence as output.

In Devlin's paper [9], extensive experiments were conducted on 800 M words BooksCorpus and 2500M words English Wikipedia with 11 NLP tasks. The resulted model is universal and would be able to adapt to various NLP tasks easily, with no need to retrain the model from scratch every single time. Specifically, it has made advancements of 7.7% on GLUE, 4.6% on MultiNLI, 1.5% on

SQuAD v1.1 QA Test F1, and 5.1% on SQuAD v2.0 Test F1. Moreover the high performance of BERT is not only limited to the general-purpose language domain; it has already been seen in other domains as well. The fine-tuned BERT models beat state-of-the-art results for many different difficulty level NLP tasks. Fig. 1 shows the basic BERT pretraining and fine-tuning procedures, and Fig. 2 illustrates the input representation flow of the BERT model.

3 Biomedical and clinical BERT

There are two steps to apply the BERT models in biomedical and clinical domain: pretraining and fine-tuning. Pretrained models normally refer to the class of models that are pretrained on certain domain-specific text corpora, while fine-tune BERT models refer to the class of models that are fine-tuned to complete some specific tasks in application domain with the adoption of pretrained models. The tasks of fine-tuned BERT models include some generic and domain-specific natural language processing tasks such as name entity recognition and relation classification, some predictions such as hospital readmission, and some classification tasks such as medication recommendations. In the past year, research conducted on this path in the medical domain has made great progress. In this section, we aim to elaborate on these novel models proposed based on BERT in the past year. Meanwhile, we will empirically compare and discuss the major contributions, architectures, applied datasets, and designed experiments of these domain-specific BERT-based models.

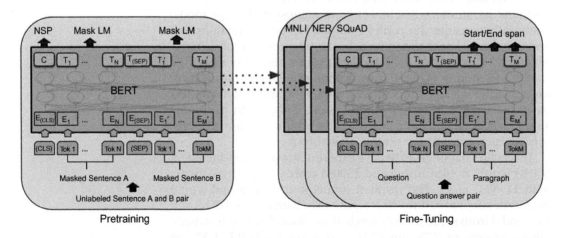

Fig. 1 Basic BERT pre-training and fine-tuning procedures [9].

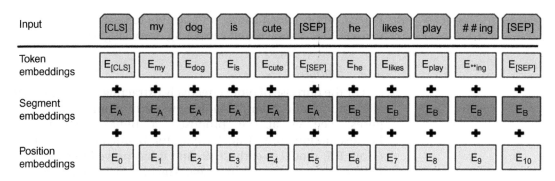

Fig. 2 BERT input representation [9].

3.1 Pretrained models

A pretrained model usually refers to the relatively general domain model built by someone else, and it is applicable to different specific domain tasks. The original BERT pretrained models were trained on Wikipedia and BooksCorpus. However, based on the studies conducted up to now, it is shown that pretrained models developed on domain-specific datasets usually provide better and more robust performance on specific tasks, such as the biomedical and clinical domain where medical jargons and professional vocabularies are common. In the past year the rise of BERT model has caught the attention of almost all researchers in biomedical and clinical NLP communities. And the number of pretrained models proposed during this period of time is significant. Among them, there are several more influential pretrained models in specific domains, which include but are not limited to BioBERT, ClinicalBERT, and SciBERT.

3.1.1 Original BERT pretrained models

The original BERT model is introduced by Google in November 2018. Since the release of the robust NLP model, there have been a hot discussion and rise of research conducted on this pretrained model. Initially in Devlin et al. [9]'s paper, they presented two different versions of the BERT model, one named BERT base and the other named BERT large. The BERT base model, which is widely applied in biomedical and clinical domain, consists of 12 layers, 768 hidden sizes, 12 self-attention heads, and a total of 110M parameters. Meanwhile the BERT large doubles the layers of BERT base with 24 layers, with 1024 hidden sizes, 16 self-attention heads, and a total of 340M parameters. The model practiced pretraining on general purpose text corpuses of Wikipedia and

BooksCorpus, which allows it to be able to be simply adopted to various NLP tasks without retraining every single time.

3.1.2 BioBERT

For the purpose of more effectively extracting salient and valuable information from the large amount of biomedical text corpuses, Lee et al. [5] proposed a Bidirectional Encoder Representation from Transformers for Biomedical Text mining (BioBERT) model in early 2019. And it has proved to outperform BERT on domain-specific datasets. The BioBERT model is theoretically built for solving various biomedical text mining problems, including classical name entity recognition (NER) problems, relation extraction (RE) problems, and questing answering (QA) problems. The major contributions of this model are as follows: (1) It is considered to be the first pretrained BERT model specifically targeting in biomedical domain; (2) it has achieved significantly better results than the existing models in the biomedical domain in specific NLP tasks, such as name entity recognition (NER) that was improved by 0.51 in F1 score, relation extraction that was improved by 3.49 in F1 score, and question answering that was improved by 9.61 in MRR score; and (3) the model requires minimum modifications in the architecture to perform competitively.

In terms of the architecture, the BioBERT uses the same structure as the basic BERT model proposed by Devlin [9]. BioBERT uses a multilayered bidirectional transformer and learns the text representations by making predictions on masked tokens and the potential next sentences. And the datasets used in this paper are mainly a various combination of different sources served to both pretrain and fine-tune the BioBERT model. Specifically the combination consists of two general purpose language datasets for pretraining on English Wikipedia (2.5 billion words) and BooksCorpus (0.8 billion words) and two biomedical domain-specific datasets for pretraining on PubMed abstracts (4.5 billion words) and PMC full text articles (13.5 billion words). The pretraining process was exercised on the BioBERT model with the combination of Wiki + Books, Wiki + Books + PubMed, Wiki + Books + PMC, and Wiki + Books + PubMed + PMC respectively. In fact, it took 200–700 k steps to pretrain the model of BioBERT, which is initialized from BERT.

3.1.3 Clinical BERT

In the clinical notes domain, there is also a popular BERT model proposed by Huang et al. [7] in early 2019. The introduced model is named ClinicalBERT, which focuses on the text corpuses of clinical notes and the application to a specific task of hospital readmission prediction. Hospital readmission has always been an ongoing hot research problem in clinical domain. The lower readmission rate is not only good for patients to save money and time but also good for the hospitals to ensure clinical resources are used effectively.

In essence the authors use the bidirectional transformers, ClinicalBERT, to develop a prediction model and to evaluate the effectiveness of learning deep representations of clinical notes. Technically the architecture of ClinicalBERT is the same as the one practiced on BERT. It takes the advantage of BERT's capability of modeling the long-term dependencies, thus be able to capture the word similarity in clinical texts more effectively and accurately. Specifically the model uses surrounding words to predict the 15% masked token and uses a binary classifier to predict the potential next sentence. The objective of this pretraining procedure is to train a set of credible clinical context-based word embeddings. The resulting function is a log likelihood function of both the masked tokens and the binary variable suggesting if the two input sentences are in consecutive order.

3.1.4 SciBERT

Similarly, SciBERT introduced by Beltagy et al. [6] is another highly related pretrained language model on the basis of BERT to address the specific domain with scientific and biomedical texts. The experiments were performed in an unsupervised pretraining manner on the datasets across a variety of scientific domains, aiming to boost the performance on fine-tuning models and specific NLP task-oriented models. The experimental results have shown the great effectiveness of the SciBERT over basic BERT models in these NLP tasks, and the SciBERT has made new records to the state-of-the-art model results.

Aside from the original BERT BaseVocab [11], this paper introduces a new 30 K sized vocabulary built on scientific text corpuses using the library of SentencePiece. Specifically the scientific corpuses use a randomly picked domain-specific dataset of 1.14 M papers from Semantic Scholar to pretrain, to fine-tune BERT, and to build the model of SciBERT. The corpus is made up with 18% computer science domain paper and 82% broad biomedical domain papers. Moreover, instead of using only the abstracts of these sources, full texts of these scientific papers were applied in the experiments.

3.2 Fine-tuning for biomedical and clinical text mining

Among the existing literature a majority of the articles take the approach of constructing their analyses on the basis of the original BERT pretrained model structure [9] with different levels of fine-tuning. Some of the key-enhanced BERT pretrained models proposed include BioBERT, ClinicalBERT, Clinical BERT Embeddings, and SciBERT. These extensive studies conducted following this path with fine-tuning have proved to obtain better results than those directly applied the original BERT (trained on Wikipedia and BooksCorpus) with fine-tuning, especially when the models are aimed to complete specific natural language processing tasks such as name entity recognition, relation extraction, and question answering.

Peng et al. [12] got the inspiration from the General Language Understanding Evaluation (GLUE) benchmark proposed by Wang et al. [13] and applied transfer learning in their study to introduce the Biomedical Language Understanding Evaluation (BLUE) benchmark in mid-2019. The BLUE benchmark aims to facilitate the researchers in biomedical domain who specialize in biomedical natural language processing especially those who adopt pretraining language representations. Peng et al. conducted extensive experiments on 10 preexisting biomedical datasets, consisting of both biomedical and clinical texts, in different sizes and different difficulty levels. Table 1 shows the details of the BLUE tasks. Specifically, for the task of sentence similarity evaluation, the paper uses MedSTS and BIOSSES as the main medical corpuses; for the name entity recognition evaluation, the paper uses BC5CDR-disease and chemical and ShARe/CLEF as the main medical corpuses; for the relation extraction evaluation, the paper uses DDI, ChemProt, and i2b2 2010 as main medical corpuses; and the document classification uses HoC following by the inference task performed on MedNLI. As a result the experiments also evaluated the respective performances of pretrained language representation models of BERT and ELMo. Their experimental results, as shown in Table 2, suggest that the BERT model pretrained on the abstracts of PubMed and the clinical notes of MIMIC-III gained the best results.

Ji et al. [14] approached the biomedical entity normalization problem by building an entity normalization architecture that fine-tunes the existing BERT-based pretrained models to improve the model performances and to advance the state-of-the-art biomedical entity normalization effectively. The architecture of the entity normalization system is made up with four major

Table 1 BLUE tasks [12].

Corpus	Train	Dev	Test	Task	Metrics	Domain	Avg sent len
MedSTS, sentence pairs	675	75	318	Sentence similarity	Pearson	Clinical	25.8
BIOSSES, sentence pairs	64	16	20	Sentence similarity	Pearson	Biomedical	22.9
BC5CDR-disease, mentions	4182	4244	4424	NER	F1	Biomedical	22.3
BC5CDR-chemical, mentions	5203	5347	5385	NER	F1	Biomedical	22.3
ShARe/CLEFE, mentions	4628	1075	5195	NER	F1	Clinical	10.6
DDI, relations	2937	1004	979	Relation extraction	Micro F1	Biomedical	41.7
ChemProt, relations	4154	2416	3458	Relation extraction	Micro F1	Biomedical	34.3
i2b2 2010, relations	3110	11	6293	Relation extraction	F1	Clinical	24.8
HoC, documents	1108	157	315	Document classification	F1	Biomedical	25.3
MedNLI, pairs	11,232	1395	1422	Inference	Accuracy	Clinical	11.9

components: a preprocessing module, a candidate concept generation module, a candidate concept ranking module, and an unlinkable mention prediction module. The detailed flow of the system is shown in Fig. 3.

The extensive experiments fine-tuned the pretrained BERT, BioBERT, and Clinical BERT models. And these experiments are considered to be preliminary study, where an entity normalization architecture is proposed by practicing fine-tuning atop the pretrained BERT, BioBERT, and ClinicalBERT models. The experiments that have shown significant improvements to the baseline models are conducted on three different datasets, namely, ShARe/CLEF, National Center for Biotechnology Information (NCBI), and TAC2017ADR. Table 3 reports the comparisons that have been made on different pretrained models. And the best fine-tuned models improve the state of the art in biomedical entity normalization by 1.17%.

Similarly, Li et al. [15] investigated the biomedical entity normalization issue using fine-tuned BERT-based models with the original structure. They collected approximately 1.5 million

Table 2 Baseline performance on BLUE task test sets [12].

Task	Metrics	State of the art	ELMo	BioBERT	Proposed model Base (P)	Base (P + M)	Large (P)	Large (P + M)
MedSTS	Pearson	83.6	68.6	84.5	84.5	**84.8**	84.6	83.2
BIOSSES	Pearson	84.8	60.2	82.7	89.3	**91.6**	86.3	75.1
BC5CDR-disease	F	84.1	83.9	85.9	**86.6**	85.4	82.9	83.8
BC5CDR-chemical	F	93.3	91.5	93	**93.5**	92.4	91.7	91.1
ShARe/CLEFE	F	70	75.6	72.8	75.4	**77.1**	72.7	74.4
DDI	F	72.9	78.9	78.8	78.1	79.4	**79.9**	76.3
ChemProt	F	64.1	66.6	71.3	72.5	69.2	**74.4**	65.1
i2b2	F	73.7	71.2	72.2	74.4	**76.4**	73.3	73.9
HoC	F	81.5	80	82.9	85.3	83.1	**87.3**	85.3
MedNLI	acc	73.5	71.4	80.5	82.2	**84**	81.5	83.8
Total			78.8	80.5	82.2	**82.3**	81.5	79.2

The best performance results among the proposed models are given in bold.

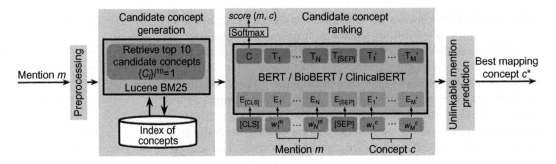

Fig. 3 The architecture of the entity normalization system [14].

electronic health records notes from the UMass Memorial Medical Center [15] and trained them in BERT to generate a novel EhrBERT pretrained model. As shown in Fig. 4, Li et al. approached the entity normalization issue as a specific task of text classification. The word representations from the transformers on the top layer is treated as the features feeding in for the normalization tasks, like BERT and BioBERT. The EhrBERT was then sevaluated on three different entity normalization corpora to examine the

Table 3 Experimental results comparison on different pretrained models [14].

	ShARe/CLEF	NCBI	TAC2017ADR
BM25	85.14	88.23	91.09
BERT$_{Base\_Cased}$	90.62	88.85	92.62
BERT$_{Base\_Uncased}$	90.58	88.65	92.97
BERT$_{Large\_Cased}$	90.73	88.85	92.87
BERT$_{Large\_Uncased}$	90.66	88.13	92.87
BioBERT$_{Base\_Cased+PubMed}$	**91.10**	88.23	**93.22**
BioBERT$_{Base\_Cased+PMC}$	90.99	88.65	92.97
BioBERT$_{Base\_Cased+PubMed+PMC}$	91.09	**89.06**	93.17
ClinicalBERT$_{Base\_Cased+MIMIC}$	90.62	88.96	92.70
ClinicalBERT$_{Large\_Cased+MIMIC}$	90.88	88.13	92.94

The best performance results among the different pretrained models on different datasets are given in bold (compare vertically).

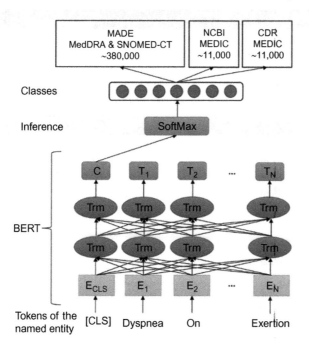

Fig. 4 Entity normalization model architecture [15].

model performances. Among the three corpora, one of them is clinical, namely, Medication, Indication, and Adverse Drug Events (MADE) corpus, and the other two are nonclinical, namely, NCBI and the Chemical-Disease Relations (CDR) corpus. And Table 4

Table 4 F1 score and standard deviation comparisons on different models [15].

Corpus and model	F1 (%), mean (SD)	Improvement compared with MetaMap or DNorm[a]
MADE[b] (gold entities[c])		
BERT[d]	67.87 (0.25)	N/A[e]
BioBERT	68.22 (0.11)	N/A
EhrBERT$_{500k}$[f]	68.74 (0.14)	N/A
EhrBERT$_{1M}$[g]	68.82 (0.29)	N/A
MADE (predicted entities[h])		
MetaMap [16]	38.59 (0)	N/A
BERT	40.81 (0.08)	+2.22
BioBERT	40.87 (0.06)	+2.28
EhrBERT$_{500k}$	40.95 (0.04)	+2.36
EhrBERT$_{1M}$	40.95 (0.07)	+2.36
NCBI[i]		
DNorm [6]	88.37 (0)	N/A
BERT	89.43 (0.99)	+1.06
EhrBERT$_{500k}$	90.00 (0.48)	+1.63
EhrBERT$_{1M}$	90.35 (1.12)	+1.98
BioBERT	90.71 (0.37)	+2.34
CDR[j]		
DNorm [6]	89.92 (0)	N/A
BERT	93.11 (0.54)	+3.19
BioBERT	93.42 (0.10)	+3.50
EhrBERT$_{500k}$	93.45 (0.09)	+3.53
EhrBERT$_{1M}$	93.82 (0.15)	+3.90

[a]DNorm: disease name normalization.
[b]MADE: Medication, Indication, and Adverse Drug Events.
[c]We used gold entity mentions as input.
[d]BERT: bidirectional encoder representations from transformers.
[e]N/A: not applicable.
[f]EhrBERT$_{500k}$: BERT-based model that was trained using 500,000 electronic health record notes.
[g]EhrBERT$_{1M}$: BERT-based model that was trained using 1 million electronic health record notes.
[h]We used MetaMap-predicted entity mentions as input.
[i]NCBI: National Center for Biotechnology Information.
[j]CDR: Chemical-Disease Relations.

presents the comparisons of F1 scores and the standard deviations on different models.

Lin et al. [17] achieved a new state-of-the-art result for the CONTAINS temporal relation extraction task by applying BERT with fine-tuning and pretraining in specific medical domain of the THYME corpus. This paper is basically a step further from the BioBERT model [5] to pretrain the model on MIMIC-III clinical datasets as well. The experimental results shown in Tables 5–7 when comparing to Lin [18] and Galvan [19], broke the state-of-

Table 5 Comparisons of CONTAINS relation on colon cancer test set [17].

Model	P	R	F1
Lin et al. [18]	0.692	0.576	0.629
Galvan et al. [19]	**0.983**	0.462	0.629
1. bi-LSTM	0.712	0.490	0.581
2. BERT	0.699	0.625	0.660
3. BERT-T	0.735	0.613	0.669
4. BERT-TS	0.670	**0.697**	0.683
5. BioBERT(pmc)-TS	0.674	0.695	**0.684**
6. BERT-MIMIC-TS	0.673	0.686	0.679

The best precision, recall, and F1 score obtained from different models are given in bold (compare vertically).

Table 6 Comparisons of CONTAINS relation on brain cancer test set [17].

Model	P	R	F1
Lin et al. [18]	**0.514**	0.585	0.547
BERT-TS	0.456	0.704	0.553
BioBERT(pmc)-TS	0.473	0.700	**0.565**
BERT-MIMIC-TS	0.457	**0.715**	0.558

The best precision, recall, and F1 score obtained from different models are given in bold (compare vertically).

Table 7 Within-sentence and cross-sentence results on colon cancer—development set [17] (https //www.aclweb.org/anthology/W19-1908).

Category	P	R	F1
Within-sentence	0.621	0.712	0.663
Cross-sentence	0.359	0.310	0.333

the-art records by improving 0.055 point for in-domain task and 0.018 point for cross-domain task.

Amin et al. [20] used transfer learning along with pretrained BERT and BioBERT model to tackle the multilabel classification of ICD-10 code problem. The text corpuses of nontechnical summaries (NTSs) in animal experiments naturally contain different architectures in the setting of multilabel classifications. The paper aims to fulfill the task of being able to assign ICD-10 codes automatically to these NTSs in animal experiments. The dataset Amin et al. use contain 8385 documents for training and 407 documents for testing in German language. These documents are translated to English by automatic translation system and then classified to one or more medical codes of the ICD-10 in a hierarchical form. After comparing with the tf-idf + linear support vector machine as baselines, shown in Tables 8 and 9, the results show considerably good achievement on F1 scores.

Xue et al. [21] investigated the entity and relation extraction problems by applying traditional BERT model to jointly learn tasks with dynamic range attention mechanism. This focused attention design empowers the BERT as a layer of shared parameter and improves the performance of generalization using feature representation, where the model architecture is shown in Fig. 5. Compared with the existing models, the proposed model structure benefits from the strong capability of BERT in feature extraction by getting better task representation vectors and less complex work on optimization. Meanwhile the model does not significantly change the original BERT structure that prior knowledge contained in the pretrained models parameters can be used directly. The experiments conducted in this study are based on real clinical data collected from Shuguang Hospital on the coronary angiography texts. These results have shown (Table 10) improvement on the state-of-the-art models by 1.65% on F1 score of the name entity recognition task and 1.22% on F1 score of the relation classification task.

Mao et al. [22] used the pretrained BERT-base model plus fine-tuning to propose a BERT-CRF model for medical document anonymization, named Hadoken. The datasets in the experiments are provided by the Medical Document Anonymization (MEDDOCAN) community challenge task who aims for setting new state-of-the-art results in both name entity recognition and entity-type classification tasks. The MEDDOCAN provides synthetic clinical cases for participants to produce the deidentification of protected health information (PHI) in the datasets. The experiments show some degree of differences in system architecture compared with many existing paper; however, as their

Table 8 Model comparisons on development set [20] (https //www.researchgate.net/profile/Saadullah_Amin2/publication/335681972_MLTDFKI_at_CLEF_eHealth_2019_Multilabel_Classification_of_ICD10_Codes_with_BERT/links/5d742a00299bf1cb809043cd/MLT-DFKI-at-CLEF-eHealth-2019-Multi-label-Classification-of-ICD-10-Codes-with-BERT.pdf), with italic as the best underline as the worst score.

Models		P	R	F$_1$
Baseline	TF-IDF$_{de}$	*90.72*	58.73	71.30
	TF-IDF$_{en}$	90.69	65.45	76.03
CNN	FT$_{de}$	86.08	57.37	68.85
	FT$_{en}$	85.76	61.59	71.69
	PubMed$_{en}$	87.95	65.10	74.82
HAN	FT$_{de}$	78.86	58.79	67.37
	FT$_{en}$	83.52	64.50	72.79
	PubMed$_{en}$	85.10	69.61	76.58
SLSTM	FT$_{de}$	85.55	64.86	73.76
	FT$_{en}$	87.53	67.65	76.32
	PubMed$_{en}$	87.33	70.09	77.77
CLSTM	FT$_{de}$	83.60	63.97	72.48
	FT$_{en}$	84.39	69.14	76.01
	PubMed$_{en}^{†}$	87.87	70.21	78.05
BERT	Multi$_{de}$	70.96	83.41	76.68
	BERT$_{en}$	79.63	84.60	82.04
	BioBERT$_{en}^{‡}$	80.35	*85.61*	*82.90*
Ensemble (†, ‡)		86.29	83.11	**84.67**

The bold is the best F1 score of the models as ensemble. The numbers with underline are the worst scores while italic are best scores.

datasets are not the major ones in biomedical domain, it is hard to make a very objective judgment on its performance.

In the model of BioBERT [5], after the pretraining process, the fine-tuning process takes the pretrained model of BioBERT and exercises it on specific NLP tasks like name entity recognition, relation extraction, and question answering, with minimum modifications of the architecture. This process of fine-tuning is actually in a similar manner as BERT. In detail the model uses WordPiece [11] tokenization proposed by Wu et al. [16] first to reduce the issue of out of vocabulary. In other words the new words appear in the texts will thus be transferred as some form of frequently seen subwords.

Table 9 Model comparisons on test set [20] (https //www.researchgate.net/profile/Saadullah_Amin2/publication/335681972_MLTDFKI_at_CLEF_eHealth_2019_Multi-label_Classification_of_ICD-10_Codes_with_BERT/links/5d742a00299bf1cb809043cd/MLT-DFKI-at-CLEF-eHealth-2019-Multi-label-Classification-of-ICD-10-Codes-with-BERT.pdf), with italic as the best underline as the worst score.

Models		Original			Modified		
		P	R	F_1	P	R	F_1
Baseline	TF-IDF$_{de}$	*89.58*	52.74	66.39	*93.01*	52.74	67.31
	TF-IDF$_{en}$	88.31	60.79	72.01	91.53	60.79	73.06
CNN	FT$_{de}$	80.30	54.66	65.04	86.99	54.66	67.13
	FT$_{en}$	78.09	58.74	67.05	83.33	58.74	68.91
	PubMed$_{en}$	80.89	64.36	71.69	86.74	64.36	73.90
HAN	FT$_{de}$	71.45	54.66	61.93	80.60	54.66	65.14
	FT$_{en}$	75.88	62.70	68.67	82.10	62.70	71.10
	PubMed$_{en}$	79.51	66.41	72.37	84.82	66.41	74.49
SLSTM	FT$_{de}$	79.17	64.11	70.85	85.37	64.11	73.23
	FT$_{en}$	82.53	65.77	73.20	86.26	65.77	74.63
	PubMed$_{en}$	77.13	68.07	72.32	83.15	68.07	74.85
CLSTM	FT$_{de}$	83.60	63.97	72.48	87.52	63.97	73.91
	FT$_{en}$	75.74	65.00	69.96	82.62	65.00	72.76
	PubMed$_{en}$†	82.15	68.19	*74.52*	86.82	68.19	76.39
BERT	Multi$_{de}$	54.10	83.39	65.62	68.23	83.39	75.05
	BERT$_{en}$	62.09	83.26	71.11	75.20	83.26	79.03
	BioBERT$_{en}$‡	63.68	*85.56*	73.02	76.57	*85.56*	*80.82*
Ensemble (†, ‡)		74.44	81.86	**77.98**	83.13	81.86	**82.49**

The best F1 score of the ensemble model is given in bold.

For NER tasks, since BioBERT learns embeddings generated from WordPiece [11] from scratch while pretraining, it uses one single layer of output based on the representations generated from the last layer to compute the probabilities of token level BIO. For relation extraction the BioBERT applies the same sentence classifier as Devlin's BERT [9], where the classification task is done by a single layer of output with the [CLS] token. The experimental results of relation extraction were then presented using anonymized name entities with predefined tags, introduced by Bhasuran et al. [23]. Lastly, for the question answering task, Lee et al. used the BERT model designed for SQuAD [24] and a single layer of output, similar to NER, to compute the answer phrases locations at token level. Tables 11–13 show the comparisons of

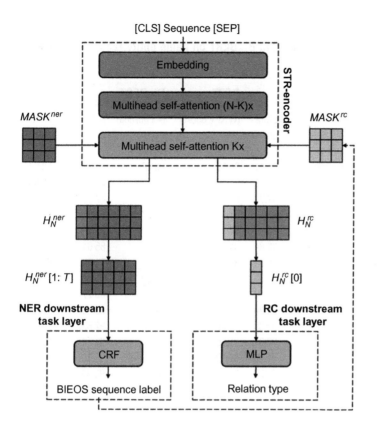

Fig. 5 Model architecture [21].

Table 10 Model comparisons on the task of joint entity and relation extraction [21].

Methods	NER			RC with predicted entities		
	Precision	Recall	F1-score	Precision	Recall	F1-score
Joint-Bi-LSTM	93.84%	96.69%	95.24%	93.64%	81.75%	87.29%
Proposed model [21]	**96.69%**	**97.09%**	**96.89%**	**95.41%**	**82.54%**	**88.51%**

The proposed model performance on different precision, recall, and F1 score measures is given in bold.

performance between BERT and BioBERT on name entity recognition task, relation extraction task, and question answering task, respectively.

Similar to BioBERT, after the pretraining is done, the Clinical-BERT [7] model proceeds to fine-tune itself for the specific clinical task of prediction on hospital readmission. The fine-tuning

Table 11 Comparison between BERT and BioBERT on name entity recognition task [5].

Entity type	Datasets	Metrics	State of the art	BERT (Wiki + Books)	BioBERT (+ PubMed)	BioBERT (+ PMC)	BioBERT (+ PubMed + PMC)
Disease	NCBI disease (Dogan et al.[25])	P	86.41	84.12	86.76	86.16	**89.04**
		R	88.31	87.19	88.02	89.48	**89.69**
		F	87.34	85.63	87.38	87.79	**89.36**
	2010 i2b2/VA (Uzuner et al. [26])	P	87.44	84.04	85.37	85.55	**87.50**
		R	**86.25**	84.08	85.64	85.72	85.44
		F	**86.84**	84.06	85.51	85.64	86.46
	BC5CDR (Li et al. [27])	P	85.61	81.97	85.80	84.67	**85.86**
		R	82.61	82.48	86.60	85.87	**87.27**
		F	84.08	82.41	86.20	85.27	**86.56**
Drug/chemical	BC5CDR (Li et al. [27])	P	**94.26**	90.94	92.52	92.46	93.27
		R	92.38	91.38	92.76	92.63	**93.61**
		F	93.31	91.16	92.64	92.54	**93.44**
	BC4CHEMD (Krallinger et al. [28])	P	91.30	91.19	91.77	91.65	**92.23**
		R	87.53	88.92	**90.77**	90.30	90.61
		F	89.37	90.04	91.26	90.97	**91.41**
Gene/protein	BC2GM (Smith et al. [29])	P	81.81	81.17	81.72	82.86	**85.16**
		R	81.57	82.42	83.38	**84.21**	83.65
		F	81.69	81.79	82.54	83.53	**84.40**
	JNLPBA (Kim et al. [30])	P	**74.43**	69.57	71.11	71.17	72.68
		R	**83.22**	81.20	83.11	82.76	83.21
		F	**78.58**	74.94	76.65	76.53	77.59
Species	LINNAEUS (Gerner et al. [31])	P	92.80	91.17	91.83	91.62	**93.84**
		R	**94.29**	84.30	84.72	85.48	86.11
		F	**93.54**	87.6	88.13	88.45	89.81
	Species-800 (Pafilis et al. [32])	P	74.34	69.35	70.60	71.54	**72.84**
		R	75.96	74.05	75.75	74.71	**77.97**
		F	74.98	71.63	73.08	73.09	**75.31**
	Average	P	85.38	82.61	84.16	84.19	**85.82**
		R	85.79	84.00	85.64	85.68	**86.40**
		F	85.53	83.25	84.82	84.87	**86.04**

Bold is the best performance models on NER task, with different datasets; underline is the second best.

Table 12 Comparison between BERT and BioBERT on relation extraction task [5].

Entity type	Datasets	Metrics	State of the art	BERT (Wiki +Books)	BioBERT (+ PubMed)	BioBERT (+ PMC)	BioBERT (+ PubMed + PMC)
Gene disease	GAD (Bravo et al. [33])	P	**79.21**	74.28	76.43	75.20	75.95
		R	**89.25**	85.11	87.65	86.15	88.08
		F	**83.93**	79.33	81.66	80.30	81.57
	EU-ADR (Van Mulligen et al. [33a])	P	76.43	75.45	78.04	**81.05**	80.92
		R	**98.01**	96.55	93.86	93.90	90.81
		F	85.34	84.71	85.22	**87.00**	85.58
Gene chemical	CHEMPROT (Krallinger et al. [34])	P	74.80	74.01	75.50	73.71	**76.63**
		R	56.00	70.79	**76.86**	75.55	76.74
		F	64.10	72.36	76.17	74.62	**76.68**
	Average	P	76.81	74.58	76.66	76.66	**77.83**
		R	81.09	84.15	**86.12**	85.20	85.21
		F	77.79	78.80	**81.02**	80.64	81.28

Bold is the best performance models on RE task, with different datasets; underline is the second best.

Table 13 Comparison between BERT and BioBERT on question answering task [5].

Datasets	Metrics	State of the art	BERT (Wiki + Books)	BioBERT (+ PubMed)	BioBERT (+ PMC)	BioBERT (+ PubMed + PMC)
BioASQ 4b (Tsatsaronis et al. [35])	S	20.59	26.75	28.96	35.59	**36.48**
	L	29.24	40.80	45.46	**52.43**	48.89
	M	24.04	32.35	34.95	**42.09**	41.05
BioASQ 5b (Tsatsaronis et al. [35])	S	41.82	37.73	**43.37**	40.45	41.56
	L	**57.43**	49.59	55.75	52.32	54.00
	M	47.73	42.50	**48.28**	44.57	46.32
BioASQ 6b (Tsatsaronis et al. [35])	S	25.12	28.45	**38.83**	37.12	35.58
	L	40.20	46.85	50.55	48.42	**51.39**
	M	29.28	35.18	**42.76**	41.15	42.51
Average	S	29.18	30.97	37.05	37.72	**37.87**
	L	42.29	45.75	50.59	51.06	**51.43**
	M	33.68	36.68	42.00	42.60	**43.29**

Bold is the best performance models on QA task, with different datasets; underline is the second best.

Table 14 Performance comparison among models and ClinicalBERT [7].

Model	Area under receiver operating characteristic	Area under precision recall	Recall at precision of 80%
ClinicalBERT	0.768 ± 0.027	0.747 ± 0.029	0.255 ± 0.113
Bag of words	0.684 ± 0.025	0.674 ± 0.027	0.217 ± 0.119
BiLSTM	0.694 ± 0.025	0.686 ± 0.029	0.223 ± 0.103

The best performance results among the models are given in bold.

involves a binary classification module to predict whether the target patient will be readmitted to the hospital within the next 30 days cycle. And as the paper presents, the fine-tuned ClinicalBERT performs significantly better than classical bag of words method and BiLSTM method, as shown in Table 14.

It is worth noting that their experiments were built on patients' discharge summaries and the clinical notes from intensive care unit (ICU) for the first few days. In fact the datasets applied in the ClinicalBERT is the same as the ones used by Alsentzer et al. [8], known as the MIMIC-III dataset proposed by Johnson et al. [36]. The returned results were confirmed to be able to capture semantic relationships accurately by physicians, shown in Table 15.

Indeed, it exceeds the predictions made by a deep language model by a 15% relative improvement in recall on 30-day hospital readmission rate with false alarms maintained at a fixed rate (Tables 16 and 17).

Table 15 ClinicalBERT capturing relationships between clinical terms accurately [7].

Model	Pearson correlation
ClinicalBERT	0.670
WORD2VEC	0.553
FastText	0.487

The best performance result among the models is given in bold.

Table 16 ClinicalBERT's prediction results with discharge summaries [7].

Model	Area under receiver operating characteristic	Area under precision recall	Recall at precision of 80%
ClinicalB$_{ERT}$	**0.768 ± 0.027**	**0.747 ± 0.029**	**0.255 ± 0.113**
Bag of words	0.684 ± 0.025	0.674 ± 0.027	0.217 ± 0.119
BiLSTM	0.694 ± 0.025	0.686 ± 0.029	0.223 ± 0.103

The best performance results among the models are given in bold.

Table 17 ClinicalBERT's prediction results with patient's clinical notes from early on [7].

Model	24–48 h	48–72 h
ClinicalB$_{ERT}$	**0.673 ± 0.041**	**0.674 ± 0.043**
Bag of words	0.648 ± 0.029	0.654 ± 0.035
BiLSTM	0.649 ± 0.044	0.656 ± 0.035
(a) Area under receiver operating characteristic		
Model	24–48 h	48–72 h
ClinicalB$_{ERT}$	**0.670 ± 0.042**	**0.677 ± 0.044**
Bag of words	0.650 ± 0.027	0.657 ± 0.026
BiLSTM	0.660 ± 0.036	0.668 ± 0.028
(b) Area under precision recall		
Model	24–48 h	48–72 h
ClinicalB$_{ERT}$	**0.167 ± 0.090**	**0.171 ± 0.107**
Bag of words	0.144 ± 0.094	0.122 ± 0.106
BiLSTM	0.143 ± 0.080	0.150 ± 0.081
(c) Recall at precision of 80%		

The best performance results among the models are given in bold.

Aside from that, Alsentzer et al. [8] aim to fill in the gap between general purpose trained BERT and the BERT on special clinical context corpora by providing two publicly available clinical BERT models and pretrained embeddings. The first part of the models is pretrained solely by clinical text, and the other is

Table 18 Task dataset evaluation metrics, output dimensionality, and train/dev/test dataset sizes [8]

Dataset	Metric	Dim	# Sentences		
			Train	Dev	Test
MedNLI	Accuracy	3	11,232	1395	1422
i2b2 2006	Exact F1	17	44,392	5547	18,095
i2b2 2010	Exact F1	7	14,504	1809	27,624
i2b2 2012	Exact F1	13	6624	820	5664
i2b2 2014	Exact F1	43	45,232	5648	32,586

fine-tuned on top of the BioBERT model. Specifically the first part of the experiment utilizes the text components in the patient's profile and trains with the BERT base model in clinical context. Then the second part of the experiment is constructed on the BioBERT model built on discharge summaries.

The clinical text datasets used in Alsentzer et al.'s two-part experiments include approximately 2 million notes from MIMIC-III v1.4 database proposed by Johnson et al. [36]. And the two Clinical BERT and Clinical BioBERT training procedures are completely standard. For all the specific tasks that the model is going to complete, including name entity recognition at single token level, deidentification task, or the begin sentence token for MedNLI, the two BERT models are simply fine-tuned before passing through a single layer of classification output. Table 18 shows the task dataset details.

In the specific tasks that we examine, as shown in Table 19, we find that our clinical embeddings can perform much better in

Table 19 Accuracy on MedNLI and Exact F1 score on i2b2 across different clinical NLP tasks [8].

Dataset	Metric	Dim	# Sentences		
			Train	Dev	Test
MedNLI	Accuracy	3	11,232	1395	1422
i2b2 2006	Exact F1	17	44,392	5547	18,095
i2b2 2010	Exact F1	7	14,504	1809	27,624
i2b2 2012	Exact F1	13	6624	820	5664
i2b2 2014	Exact F1	43	45,232	5648	32,586

deidentification task than those pretrained in general domain and those pretrained with specific biomedical corpora, BioBERT. Furthermore the ones trained with clinical notes text corpuses enjoy more selective benefits in performance compared with other settings.

The experiments of SciBERT [6] were constructed on the following natural language processing tasks as many other articles in this domain: name entity recognition (NER), Participant Intervention Comparison Outcome (PICO) extraction, text classification (CLS), relation classification (REL), and dependency parsing (DEP). Among the five major tasks, NER and PICO are sequence labeling tasks, CLS and REL are classification tasks, and DEP is prediction task on token dependencies within the sentence. Following that categorization the paper uses different datasets to complete different tasks, with each at a size of 12 K training samples, as follows: EBM-NLP proposed by Nye et al. [37] was used for PICO spans in abstracts from clinical trial, SciERC proposed by Luan et al. [38] was used for annotating the relationships and entities in the abstracts of computer science, ACL-ARC proposed by Jurgens et al. [39] and SciCite proposed by Cohan et al. [40] gives sentences intention labels where these papers make citations to other scientific papers, and Paper Field proposed by Sinha et al. [41] to link paper titles to one of the given seven scientific study fields.

The architecture of the SciBERT follows the basic BERT base structure, where it is in essence a multibidirectional transformer, and it learns text representations by making predictions on masked tokens and potential next sentences. In total the SciBERT was trained in four versions, namely, cased for NER tasks [9] or uncased for all other tasks [9] and on BaseVocab or on SciVocab. The two versions built on BaseVocab would be undergoing the fine-tuning process from BERT-Base model. And the two versions built on SciVocab would be built and trained from scratch. The initial maximum length of sentences are set to be 128 tokens until the training loss cease to decrease, and then the maximum length of sentences are set to 512 tokens afterward.

For the fine-tuning process, SciBERT [6] basically used same choices on architecture, optimization, and hyperparameters as Devlin et al. [9] on BERT. The only differences are the application of additional conditional random field (CRF) to sequence labeling process and the usage of Dozat and Manning's model [42] for DEP. Besides, Beltagy et al. set the dropout rate equal to 0.1 and use Adam [43] to optimize the loss of cross entropy. Furthermore the learning rate of 5×10^{-6}, 1×10^{-5}, 2×10^{-5}, or 5×10^{-5} with a slanted triangular schedule [44] and a batch size of 32 were applied to fine-tune the two to five epochs.

Meanwhile the paper also managed to train specific task-oriented models on top of the frozen BERT embeddings. In the sequence labeling tasks, the paper uses a BiLSTM layers and CRF to ensure the well-formed predictions. In the text classification tasks, the paper sends each BERT vectored sentence into a multilayered perceptron on the concatenation of first and last BiLSTM vectors. In the DEP task, Dozat and Manning's full model [42] was applied. And finally the Adam was used for optimizing the loss of cross entropy, with original BERT weights and a dropout rate of 0.5. Table 20 shows the test performances of all BERT variants on all tasks and datasets.

When comparing the results generated by SciBERT with BioBERT, we can see that SciBERT and BioBERT are comparable and performing well on different datasets. Table 21 is the comparison between SciBERT with BioBERT on biomedical datasets, Table 22 shows the different tokenization methods adopted among the models, and Table 23 shows different data combinations across these comparable models.

Table 20 Test performances of all BERT variants on all tasks and datasets [6].

Field	Task	Dataset	SOTA	BERT-Base Frozen	BERT-Base Fine-tune	SciBERT Frozen	SciBERT Fine-tune
Bio	NER	BC5CDR (Li et al. [27])	88.85[7]	85.08	86.72	88.73	**90.01**
		JNLPBA (Collier and Kim [45])	**78.58**	74.05	76.09	75.77	77.28
		NCBI-disease (Dogan et al. [25])	**89.36**	84.06	86.88	86.39	88.57
	PICO	EBM-NLP (Nye et al. [37])	66.30	61.44	71.53	68.30	**72.28**
	DEP	GENIA (Kim et al. [46])—LAS	**91.92**	90.22	90.33	90.36	90.43
		GENIA (Kim et al. [46])—UAS	**92.84**	91.84	91.89	92.00	91.99
	REL	ChemProt (Kringelum et al. [47])	76.68	68.21	79.14	75.03	**83.64**
CS	NER	SciERC (Luan et al. [38])	64.20	63.58	65.24	65.77	**67.57**
	REL	SciERC (Luan et al. [38])	n/a	72.74	78.71	75.25	**79.97**
	CLS	ACL-ARC (Jurgens et al. [39])	67.9	62.04	63.91	60.74	**70.98**
Multi	CLS	Paper Field	n/a	63.64	65.37	64.38	**65.71**
		SciCite (Cohan et al. [40])	84.0	84.31	84.85	**85.42**	85.49
Average				73.58	77.16	76.01	79.27

The best performance results among the models on different fields, tasks, and datasets are given in bold.

Table 21 Comparisons between SciBERT with BioBERT on biomedical datasets [6].

Task	Dataset	BioBert	SciBert
NER	BC5CDR	88.85	90.01
	JNLPBA	77.59	77.28
	NCBI-disease	89.36	88.57
REL	ChemProt	76.68	83.64

Table 22 Different tokenization methods adopted across different models (https // towardsdatascience.com/how-to-apply-bert-in-scientific-domain-2d9db0480bd9).

BERT Tokenizer	SciBERT Tokenizer	BioBERT Tokenizer
WordPiece: token separation	ScispaCy: splitting document to sentence. SentencePiece: token separation	WordPiece: token separation
Different tokenization method among models		

Table 23 Some of the data combinations across different models.

BERT (# of tokens)	SciBERT (# of tokens)	BioBERT (# of tokens)
English Wikipedia: 2.5B BooksCorpus: 0.8B	Biomedical paper: 2.5B Computer Science paper: 0.6B	English Wikipedia: 2.5B BooksCorpus: 0.8B PubMed Abstracts: 4.5B PMC full text: 13.5B

3.3 Discussions

As we can see, the multiple self-attention transformers models have become more popular and robust in various research fields, including biomedical and clinical domain, after the release of BERT. BERT indeed can offer promising results not only to keep up with the state-of-the-art results but also to make new state-of-the-art records. The aforementioned sections have already given a general picture of how both the original BERT-based and the enhanced BERT-based pretrained models perform in the medical domain. The implication is that a universal pretrained BERT model that is able to capture salient features and achieve the state-of-the-art results in all different NLP-specific tasks may not exist. To get the best state-of-the-art experimental results, it is necessary to add in extra fine-tuning steps on domain-specific datasets.

As a result, there is still great room for further improvement of BERT model in the medical and clinical domain. For example, in Alsentzer et al.'s Clinical BERT embedding paper [8], they use a signal layer of classification model to do evaluation on the experimental results. This can be considered to be a good start or a preliminary study for further studies. The paper has already proposed the expectation on BERT model to practice self-learning on the context and the content of the clinical text corpuses. In the near future, one can expect some other scholars to follow this path by adopting other more advanced architectures of models to provide a better and also more comprehensive experimental result.

On the other hand, as BERT is performing more robust and well in short inputs, it is necessary for the NLP community to think of a better approach to solve the long input problem. For example, in Huang et al.'s Clinical BERT paper [7], they proposed to use some mathematical tricks to solve the problem brought by long clinical notes input. As the nature of clinical notes may contain long sentence inputs, the BERT model has a limitation on the maximum length of an input sequence; there would be a potential significant improvement in the currently existing models if the problem is solved.

4 Conclusions and future work

In this chapter, we generally give an overview of some significant BERT-based models released in the medical domain in the past year. There are two steps to apply the BERT models in biomedical and clinical domain: pretraining and fine-tuning. According to the pretrained model they used, the existing BERT models can be classified into two categories, namely, the original BERT-based models and the enhanced BERT-based models. In general, these existing BERT models in both approaches have proved to be effective in various biomedical and clinical NLP tasks, which include but are not limited to name entity recognition, relation extraction, and question answering. However, there are still certain problems exist, such as the way of dealing with long information inputs.

To further improve and enhance the performance of these models, we believe it is necessary to begin with the better exploration, investigation, and understanding of the existing problems in adopting BERT-based models to the medical domain and its closely related fields [48–51], such as its limitations in nature. Then, we can slowly move toward pretraining more novel BERT-based enhanced models, which can tailor better to these limitations and fine-tune on domain-specific datasets. Meanwhile the continuation of exploring the potential of BERT-based models is also necessary. The NLP research community should continue to fine-tune the existing NLP tasks, such as name entity recognition, which have already shown promising results. More other domain-specific diseases, applications, and datasets should also be brought into the BERT studies [52–63]. Last but not least, by the time this chapter is written, other transformer-based models such as RoBERTa [64] and XLNet [65] are also released with incredibly good and robust performance. It is worth trying to apply these novel approaches to the biomedical and clinical studies in the near future.

Acknowledgment

This work is supported by the Natural Sciences and Engineering Research Council (NSERC) of Canada, an NSERC CREATE award in ADERSIM (http://www.yorku.ca/adersim) the York Research Chairs (YRC) program, and an ORF-RE (Ontario Research Fund-Research Excellence) award in BRAIN Alliance (http://brainalliance.ca).

References

[1] B. Chiu, G. Crichton, A. Korhonen, S. Pyysalo, How to train good word embeddings for biomedical NLP, in: Proceedings of BioNLP Workshop, 2016, pp. 166–174.
[2] Q. Chen, Y. Peng, L. Zhiyong, BioSentVec: creating sentence embeddings for biomedical texts, in: Proceedings of the 7th IEEE International Conference on Healthcare Informatics, 2019.
[3] M.E. Peters, W. Ammar, C. Bhagavatula, R. Power. Semi-supervised sequence tagging with bidirectional language models. In Proceedings of ACL, pages 1756–1765. 2017.
[4] N.R. Smalheiser, A.M. Cohen, G. Bonifield, Unsupervised low-dimensional vector representations for words, phrases and text that are transparent, scalable, and produce similarity metrics that are not redundant with neural embeddings, J. Biomed. Inform. 90 (2019) 103096.
[5] J. Lee, W. Yoon, S. Kim, D. Kim, S. Kim, C. So, J. Kang, BioBERT: a pre-trained biomedical language representation model for biomedical text mining, ArXiv (2020), https://arxiv.org/pdf/1901.08746.pdf.

[6] I. Beltagy, K. Lo, A. Cohan, SCIBERT: a pretrained language model for scientific text, ArXiv (2019), https://arxiv.org/pdf/1903.10676.pdf.
[7] K. Huang, J. Altosaar, R. Ranganath, ClinicalBert: modeling clinical notes and predicting hospital readmission, ArXiv (2019), https://arxiv.org/pdf/1904.05342.pdf.
[8] E. Alsentzer, J. Murphy, W. Boag, W. Weng, D. Jin, T. Naumann, M. McDermott, Publicly available clinical BERT embeddings, ArXiv (2019), https://arxiv.org/pdf/1904.03323.pdf.
[9] J. Devlin, M. Chang, K. Lee, K. Toutanova. BERT: pre-training of deep bidirectional transformers for language understanding. *ArXiv* 2019, https://arxiv.org/pdf/1810.04805.pdf.
[10] A. Vaswani, N. Shazeer, N. Parmar, J. Uszkoreit, L. Jones, A.N. Gomez, L. Kaiser, I. Polosukhin, Attention is all you need, in: Advances in Neural Information Processing Systems, 2017, pp. 6000–6010.
[11] Y. Wu, M. Schuster, Z. Chen, Q.V. Le, M. Norouzi, W. Macherey, M. Krikun, Y. Cao, Q. Gao, J. Klingner, A. Shah, M. Johnson, X. Liu, L. Kaiser, S. Gouws, Y. Kato, T. Kudo, H. Kazawa, K. Stevens, G. Kurian, N. Patil, W. Wang, C. Young, J. Smith, J. Riesa, A. Rudnick, O. Vinyals, G.S. Corrado, M. Hughes, J. Dean, Google's Neural Machine Translation System: Bridging the Gap Between Human and Machine Translation, abs/1609.08144(2016).
[12] Y. Peng, S. Yan, Z. Lu, Transfer learning in biomedical natural language processing: an evaluation of BERT and ELMo on ten benchmarking datasets, ArXiv (2019), https://arxiv.org/pdf/1906.05474.pdf.
[13] A. Wang, A. Singh, J. Michael, F. Hill, O. Levy, S.R. Bowman GLUE: a multi-task benchmark and analysis platform for natural language understanding. *ArXiv* 2018, https://arxiv.org/pdf/1804.07461.pdf.
[14] Z. Ji, Q. Wei, H. Xu, BERT-based ranking for biomedical entity normalization. *ArXiv* 2019, https://arxiv.org/pdf/1908.03548.pdf
[15] F. Li, Y. Jin, W. Liu, B. Rawat, P. Cai, H. Yu, Fine-tuning bidirectional encoder representations from transformers (BERT)-based models on large-scale electronic health record notes: an empirical study, JMIR Med. Inform. 7 (3) (2019) e14830. https://medinform.jmir.org/2019/3/e14830/.
[16] Y. Wu, et al., Google's neural machine translation system: bridging the gap between human and machine translation, arXiv (2016) preprint arXiv:1609.08144.
[17] C. Lin, T. Miller, D. Dligach, S. Bethard, G. Savova, A BERT-based universal model for both within- and cross-sentence clinical temporal relation extraction, in: Proceedings of the 2nd Clinical Natural Language Processing Workshop, pages 65–71 Minneapolis, Minnesota, June 7, 2019. https://www.aclweb.org/anthology/W19-1908.
[18] C. Lin, T. Miller, D. Dligach, H. Amiri, S. Bethard, G. Savova, Self-training improves recurrent neural networks performance for temporal relation extraction, in: Proceedings of the Ninth International Workshop on Health Text Mining and Information Analysis, 2018, pp. 165–176.
[19] D. Galvan, N. Okazaki, K. Matsuda, K. Inui, Investigating the challenges of temporal relation extraction from clinical text, in: Proceedings of the Ninth International Workshop on Health Text Mining and Information Analysis, 2018, pp. 55–64.
[20] S. Amin, G. Neumann, K. Dunfield, A. Vechkaeva, K.A. Chapman, M.K. Wixted, MLT-DFKI at CLEF eHealth 2019: Multi-Label Classification of ICD-10 Codes With BERT. https://www.researchgate.net/profile/Saadullah_Amin2/publication/335681972_MLT-DFKI_at_CLEF_eHealth_2019_Multi-label_Classification_of_ICD-10_Codes_with_BERT/links/5d742a00299bf1cb809043cd/MLT-DFKI-at-CLEF-eHealth-2019-Multi-label-Classification-of-ICD-10-Codes-with-BERT.pdf, 2019.

[21] K. Xue, Y. Zhou, Z. Ma, T. Ruan, H. Zhang, P. He, Fine-tuning BERT for joint entity and relation extraction in Chinese medical text, ArXiv (2019), https://arxiv.org/pdf/1908.07721.pdf.

[22] J. Mao, W. Liu, Hadoken: A BERT-CRF Model for Medical Document Anonymization, http://ceur-ws.org/Vol-2421/MEDDOCAN_paper_11.pdf, 2019.

[23] B. Bhasuran, J. Natarajan, Automatic extraction of gene- disease associations from literature using joint ensemble learning, PloS One 13 (7) (2018) e0200699.

[24] P. Rajpurkar, et al., Squad: 100,000+ questions for machine comprehension of text, arXiv (2016) preprint arXiv:1606.05250.

[25] R.I. Dogan, R. Leaman, Z. Lu, NCBI disease corpus: a resource for dis- ease name recognition and concept normalization, J. Biomed. Inform. 47 (2014) 1–10.

[26] Ö. Uzuner, et al., 2010 i2b2/va challenge on concepts, assertions, and relations in clinical text, J. Am. Med. Inform. Assoc. 18 (5) (2011) 552–556.

[27] J. Li, Y. Sun, R.J. Johnson, D. Sciaky, C. Wei, R. Leaman, A.P. Davis, C.J. Mattingly, T.C. Wiegers, Z. Lu, BioCreative V CDR task corpus: a resource for chemical disease relation extraction. Database (Oxford) (2016). https://doi.org/10.1093/database/baw068 baw068.

[28] M. Krallinger, O. Rabal, F. Leitner, et al., The CHEMDNER corpus of chemicals and drugs and its annotation principles, J. Cheminform. 7 (2015) S2. https://doi.org/10.1186/1758-2946-7-S1-S2.

[29] L. Smith, et al., Overview of biocreative ii gene mention recognition, Genome Biol. 9 (2) (2008) S2.

[30] Kim, J.-D. et al. Introduction to the bio-entity recognition task at jnlpba. In Proceedings of the international joint workshop on natural language processing in biomedicine and its applications, pages 70–75. Association for Computational Linguistics. 2004.

[31] M. Gerner, et al., Linnaeus: a species name identification system for biomedical literature, BMC Bioinform. 11 (1) (2010) 85.

[32] E. Pafilis, et al., The species and organisms resources for fast and accurate identification of taxonomic names in text, PLoS One 8 (6) (2013) e65390.

[33] A. Bravo, et al., Extraction of relations between genes and diseases from text and large-scale data analysis: implications for translational research, BMC Bioinform. 16 (1) (2015) 55.

[33a] E.M. Van Mulligen, A. Fourrier-Reglat, D. Gurwitz, M. Molokhia, A. Nieto, G. Trifiro, J.A. Kors, L.I. Furlong, The EU-ADR corpus: annotated drugs, diseases, targets, and their relationships. J. Biomed. Inform. 45 (5) (2012) 879–884, https://doi.org/10.1016/j.jbi.2012.04.004.

[34] M. Krallinger, et al., Overview of the Biocreative vi Chemical Protein Interaction Track, Proceedings of the BioCreative VI Workshop. Bethesda, MD, USA, 2017, pp. 141–146.

[35] G. Tsatsaronis, et al., An overview of the bioasq large-scale biomedical semantic indexing and question answering competition, BMC Bioinform. 16 (1) (2015) 138.

[36] A. Johnson, T. Pollard, L. Shen, et al., MIMIC-III, a freely accessible critical care database, Sci. Data 3 (2016) 160035. https://doi.org/10.1038/sdata.2016.35.

[37] B. Nye, J.J. Li, R. Patel, Y. Yang, I.J. Marshall, A. Nenkova, B.C. Wallace, A corpus with multi-level annotations of patients, interventions and outcomes to support language processing for medical literature, ACL, (2018).

[38] Y. Luan, L. He, M. Ostendorf, H. Hajishirzi, Multi-task identification of entities, relations, and coreference for scientific knowledge graph construction, EMNLP, (2018).

[39] D. Jurgens, S. Kumar, R. Hoover, D.A. McFarland, D. Jurafsky, Measuring the evolution of a scientific field through citation frames, TACL 06 (2018) 391–406.

[40] A. Cohan, W. Ammar, M. Zuylen, F. Cady, Structural scaffolds for citation intent classification in scientific publications, in: NAACL-HLT, Association for Computational Linguistics, Minneapolis, Minnesota, 2019, pp. 3586–3596.
[41] A. Sinha, Z. Shen, Y. Song, H. Ma, D. Eide, B.P. Hsu, K. Wang, An overview of microsoft academic service (MAS) and applications, WWW, (2015).
[42] T. Dozat, C.D. Manning, Deep biaffine attention for neural dependency parsing, ICLR, (2017).
[43] D.P. Kingma, J. Ba, Adam: a method for stochastic optimization, ICLR, (2015).
[44] J. Howard, S. Ruder, Universal language model fine-tuning for text classification, in: ACL, 2018.
[45] N. Collier, J. Kim, Introduction to the bio-entity recognition task at jnlpba, in: NLP-BA/BioNLP, 2004.
[46] J. Kim, T. Ohta, Y. Tateisi, J. Tsujii, GENIA corpus—a semantically annotated corpus for bio-textmining, Bioinformatics 19 (2003) i180i182.
[47] J. Kringelum, S.K. Kjærulff, S. Brunak, O. Lund, T.I. Oprea, O. Taboureau, ChemProt-3.0: a global chemical biology dis- eases mapping, Database, (2016).
[48] X. Huang, M. Zhong, L. Si, York University at TREC 2005: genomics track, in: Proceedings of the Fourteenth Text REtrieval Conference, TREC 2005, Gaithersburg, Maryland, USA, November 15-18, National Institute of Standards and Technology (NIST), 2005.
[49] X. Huang, Q. Hu, A bayesian learning approach to promoting diversity in ranking for biomedical information retrieval, in: Proceedings of the 32nd Annual International ACM SIGIR Conference on Research and Development in Information Retrieval, SIGIR 2009, Boston, MA, USA, July 19-23, 2009.
[50] X. Yin, X. Huang, Z. Li, X. Zhou, A survival modeling approach to biomedical search result diversification using Wikipedia, IEEE Trans. Knowl. Data Eng. 25 (6) (2013) 1201–1212.
[51] Z. Liang, G. Zhang, X. Huang, Q. Hu, Deep learning for healthcare decision making with EMRs, in: 2014 IEEE International Conference on Bioinformatics and Biomedicine, BIBM 2014, Belfast, United Kingdom, November 2–5, 2014, IEEE Computer Society, 2014, pp. 556–559.
[52] Z. Liang, A. Powell, I. Ersoy, M. Poostchi, K. Silamut, K. Palaniappan, P. Guo, M.A. Hossain, S.K. Antani, R.J. Maude, X. Huang, S. Jaeger, G.R. Thoma, CNN-based image analysis for malaria diagnosis, in: IEEE International Conference on Bioinformatics and Biomedicine, BIBM 2016, Shenzhen, China, December 15–18, 2016, IEEE Computer Society, 2016, pp. 493–496.
[53] Y. Liu, X. Huang, A. An, X. Yu. ARSA: a sentiment-aware model for predicting sales performance using blogs. In Proceedings of the 30th Annual International ACM SIGIR Conference on Research and Development in Information Retrieval, Amsterdam, The Netherlands, July 23–27, 2007. Pages 607–614.
[54] Y. Liu, X. Huang, A. An, X. Yu, Modeling and predicting the helpfulness of online reviews, in: Proceedings of the 8th IEEE International Conference on Data Mining (ICDM 2008), December 15–19, 2008, Pisa, Italy, IEEE Computer Society, 2008, pp. 443–452.
[55] X. Huang, F. Peng, D. Schuurmans, N. Cercone, S.E. Robertson, Applying machine learning to text segmentation for information retrieval, J. Inform. Retr. 6 (3–4) (2003) 333–362.
[56] X. Yu, Y. Liu, X. Huang, A. An, Mining online reviews for predicting sales performance: a case study in the movie domain, J. IEEE Trans. Knowl. Data Eng. 24 (4) (2012) 720–734.
[57] W. Feng, Q. Zhang, G. Hu, X. Huang, Mining network data for intrusion detection through combining SVMs with ant colony networks, J. Future Gener. Comput. Syst. 37 (2014) 127–140.

[58] Y. Liu, A. An, X. Huang, Boosting prediction accuracy on imbalanced datasets with SVM ensembles, in: Advances in Knowledge Discovery and Data Mining, 10th Pacific-Asia Conference, PAKDD 2006, Singapore, April 9–12, 2006, Series Lecture Notes in Computer Science, vol. 3918, Springer, 2006.

[59] J. Zhao, X. Huang, B. He, CRTER: using cross terms to enhance probabilistic information retrieval, in: Proceeding of the 34th International ACM SIGIR Conference on Research and Development in Information Retrieval, SIGIR 2011, Beijing, China, July 25-29, 2011, pp. 155–164.

[60] C. Tian, Y. Zhao, L. Ren, A Chinese event relation extraction model based on BERT, in: 2019 2nd International Conference on Artificial Intelligence and Big Data (ICAIBD), Chengdu, China, 2019, pp. 271–276.

[61] Z. Song, Y. Xie, W. Huang, H. Wang, Classification of Traditional Chinese Medicine Cases Based on Character-Level BERT and Deep Learning, 2019 IEEE 8th Joint International Information Technology and Artificial Intelligence Conference (ITAIC), Chongqing, China, 2019, pp. 1383–1387.

[62] S.S. Roy, V.M. Viswanatham, Classifying spam emails using artificial intelligent techniques, Int. J. Eng. Res. Afr. 22 (2016) 152–161 Trans Tech Publications.

[63] P. Samui, S.S. Roy, Balas, V. E. (Eds.)., Handbook of Neural Computation, Academic Press, 2017.

[64] Y. Liu, M. Ott, N. Goyal, J. Du, M. Joshi, D. Chen, O. Levy, M. Lewis, L. Zettlemoyer, V. Stoyanov, RoBERTa: a robustly optimized BERT pretraining approach, ArXiv (2019), https://arxiv.org/abs/1907.11692.

[65] Z. Yang, Z. Dai, Y. Yang, J. Carbonell, R. Salakhutdinov, Q. V. Le. XLNet: generalized autoregressive pretraining for language understanding. ArXiv: https://arxiv.org/pdf/1906.08237.pdf . 2019.

6

Classifying CT scan images based on contrast material and age of a person: ConvNets approach

Soumik Mitra, Sanjiban Sekhar Roy, and Kathiravan Srinivasan
Vellore Institute of Technology, Vellore, India

1 Introduction

During a computed tomography (CT) scan, a person is exposed for a short period of time to radiation through the process of ionization. The extent of radiation in a CT scan is greater than that exposed during X-ray since CT scans give more detailed information about the subject in the study. The small amount of doses of radiation emitted in a CT scan has not been seen to cause long-term harm, although at a higher amount of doses, there can be a minute increase in the risk of having chronic diseases like cancer [1]. CT scans are often administered with or without a contrast media. The contrast material improves the ability of a radiologist to view the inside images of the body in a more detailed fashion [2]. The contrast material is a special dye that tends to block X-rays and appears white on images; it helps in emphasizing the images under study [3]. A practitioner can administer the contrast material either through mouth, injections, or enema [4]. Sixteen years ago in a study, it was found about the possibility of a risk of having cancer induced by pediatric controlled computed tomography scans and the potential of having harmful effects of contrast-related exposures [5]. The study predicted risks of fatal cancer ranging from 1 per 10,000 to 1 per 1000 scanned patient, depending on their age and scanned body part. Our study provides a classification model based on the age of a patient and the contrast used in the CT scans and classifying the images for abnormalities and misclassifications to help doctors diagnose a patient based on the classified output concerning the features. We used a convolutional neural network to classify such images provided in the dataset. The dataset provided by The Cancer Imaging Archive (TCIA) includes a set of hundred CT images tagged with the age of a person and the contrast material usage. The model is successful in providing a classified result output, and it was seen that it successfully classified the images by age and contrast with an 89.99% accuracy rate [6, 7].

The remaining part of the paper has been organized as follows. The objectives of this research work have been mentioned in Section 2. Section 3 describes the basics of convolutional neural network. Section 4 resents the related research that has been carried on applications of CNN on medical images. Sections 5 and 6 present about the detail of dataset and methodology used for classification, respectively. Section 7 presents the results and discussion. Finally, Section 8 provides the conclusion of this work.

2 Objectives

1. To classify the CT scan images based on age and contrast material using CNN.
2. To find outliers and misclassifications in the images provided, for example, if the images are correctly tagged or if some of the tags based on age and contrast is missing or not. Hence, preprocessing has been accomplished on the set of images to obtain the relationships and to find input streams that are not contributing to the model.
3. To use a two-dimensional convolutional neural network to make machines learn from the CT scan images and use the 2-D CNN model to perform classifications and check if the images are correctly tagged according to the age and contrast material used.
4. To apply the model on new test CT scan images of patients to classify the scanned images that can help doctors get an idea beforehand about the probable age of a person and if the contrast being used can affect the image, it can hence further help in diagnosing a patient and see trends in those images.
5. Check for high variance or bias in CT scan images based on how the age and contrast of a person are classified.

3 Convolutional neural networks

Deep learning is a facet of artificial intelligence that is apprehensive about emulating the learning approach that men or women use to acquire specific kinds of knowledge. Deep learning model builds higher level attributes from lower level ones and hence helps in automating the process of feature construction [8, 9]. One of these models is the renowned convolutional neural network (CNN) [10–13]. A CNN consists of a set of algorithms in machine learning that can be considered as layers of processing that involves linear and nonlinear learning operators capable in learning and constructing high-level automated systems [14, 15]. A CNN uses a forward feeding artificial neural network to reproduce various patterns of multilayer perceptron (MLP) after applying to process the images such as convolution and pooling, hence making it widely popular when it comes to recognizing images and video processing. CNNs have been seen to be hugely useful when it came to classifying image content. CNNs have been seen to achieve cutting-edge performance on object recognition tasks when trained with specific values for regularization parameters. CNNs are expressed theoretically with the use of mathematics and signal processing. Our proposed method uses the training image data, expressed as x^i is of three dimensions and y^i is the label vector indicating the class of x^i, the features of an image termed as feature map, will be learned based on the equation:

$$argmin_{u_1,...,u_{loss}} \frac{1}{n} \sum_{i=1}^{n} loss(f(x^i; u_1, ..., u_{loss}), y^i) \qquad (1)$$

Further, to determine the feature masks in a 2-D CNN, operations need to be performed. In convolutional layer the convolution operation is done along with pooling layers to select features of images from neighboring pixels acquired from previous processing layers. A bias is applied to the function, and the result is passed through a hyperbolic tangential function to find the feature value at any position on the jth feature map filter in the ith layer of processing:

$$r_{ij}^{xy} = \tanh\left(b_{ij} + \sum_{m} \sum_{p=0}^{P_i-1} \sum_{q=0}^{Q_i-1} u_{ijm}^{pq} \times r_{(i-1)m}^{(x+p)(y+q)}\right) \qquad (2)$$

where m is the variable over the set of feature filters in the $(i-1)$th layer of processing, b_{ij} is the bias for the feature filter f, $u_{ij}k_{pq}$ is the value of the kernel function at position (p, q) connected to the kth feature filter, (p, q) is the two-dimensional position of the kernel function, and P_i and Q_i are the height and width of the kernel used.

3.1 Abbreviations and acronyms

See Table 1.

Table 1 Units of convolutional neural networks.

Symbol	Quantity
(x^i, y^i)	Training data
x^i	Image
y^i	Indicated label vector
u^i	Weight vectors
$loss$	The loss function
f	Class vector scores and classifier
r_{ij}^{xy}	Feature value of jth feature map in ith layer at (x, y) position
b_{ij}	Bias for feature map
α	Learning rate
Q	Function to sustain minimum loss function
$\hat{Q}$	Ground truth parameter
P_i	Height of kernel
Q_i	Width of kernel

4 Related study

Papers	Application	Method used
Zhang et al. [16]	CNN used for segmentation of tissues	Two-dimensional convolutional neural networks on multimodal input
Kleesiek et al. [17]	CNN used for brain tumor extraction	Three-dimensional convolutional neural network on multimodal input
Simonovsky et al. [18]	CNN used for similarity measurement between moving images	Three-dimensional convolutional neural network is estimating similarities between the reference and moving images stacked in the input
Wu et al. [19]	Three-dimensional ResNeXt implementation	Three-dimensional ResNeXt network using feature fusion and label smoothing for hyperspectral image classification.

Papers	Application	Method used
Shi et al. [20]	ANN used for AD and other disease classification	Multimodal stacked deep polynomial networks with an SVM classifier on top using MRI and PET
Suk et al. [21]	The restricted Boltzmann machine used for the AD, HC, and other related disease classification	Deep restricted Boltzmann machines on MRI images and PET scans
Payan and Montana [22]	CNN used for AD and other disease classification	Three-dimensional pretrained CNN model with sparse autoencoders used
Plis et al. [23]	Deep belief networks used for classification of schizophrenia and Huntington's disease	Deep belief networks evaluated on brain network estimation
Brosch and Tam [24]	Deep belief networks used for the AD and HC disease classification	Deep belief networks with convolutional restricted Boltzmann machines for manifold based learning
Huang et al. [25]	Restricted Boltzmann machine used for fMRI image separation of blind sources	RBM for both internal and functional interaction-induced latent source detection

5 Dataset

The results shown in the succeeding text are in whole or part based upon data generated by the TCGA Research Network. It contains a total number of 100 images of size 512×512 (https://wiki.cancerimagingarchive.net/display/Public/TCGA-LUAD). This dataset is free to use (https://creativecommons.org/licenses/by/3.0/). The training set and test set were divided in a 90:10 ratio. It also contains a list of patient's age and contrast

Fig. 1 The CT scan images according to the dataset given.

used for CT scan and mapped it with the images. The dataset contains eight columns—a serial number, age, contrast, contrast tag, the input path, the ID, image name, and the DICOM image name. It is seen that half of the images have contrasts applied and others do not. The contrast tag tells us what kind of contrast used if applied. An example of how the images in the dataset look like is given in Fig. 1 [26].

6 Methodology

6.1 Exploratory data analysis

To explore more about the features and how it contributes to the learning parameters, exploratory data analysis is performed. We found how age changed with contrast types and plotted it against the kernel density points through a distribution plot. This plot is shown in Fig. 2 [3].

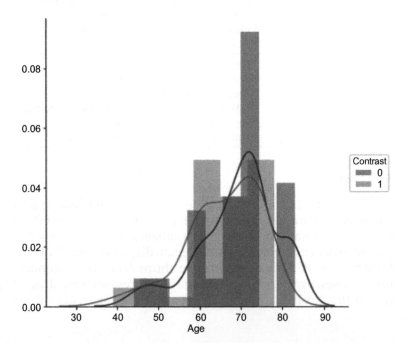

Fig. 2 Distribution between the age of a person and contrast types based on the kernel density points.

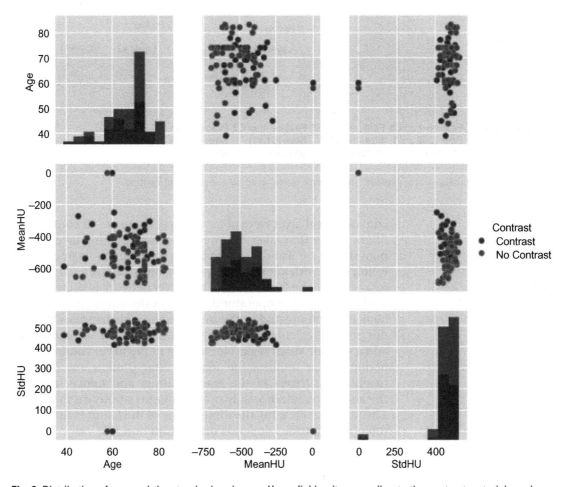

Fig. 3 Distribution of age and the standard and mean Hounsfield units according to the contrast material used.

Then, more visualization plots were created to check for variations in age and standard and mean Hounsfield unit (HU) with or without contrast; these visualizations were crucial in the analysis of the dataset to create an intuition toward the variation of each feature toward the classification and is given in Fig. 3.

6.2 Implementation of 2-D CNNs

In 2-D, all slices are cropped and normalized from the 512×512 size to 256×256 pixels [27]. The layers include two convolutional layers along with two max-pooling layers followed by the flattening layer and the artificial neural network that follows with a softmax activation function. The strides for the

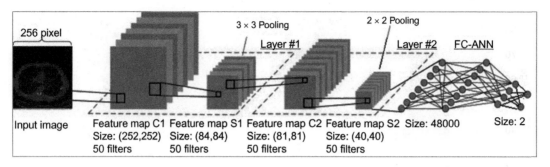

Fig. 4 The convolutional neural network architecture model used by our proposed method.

convolutional layer were of 5 × 5 sizes. There are 50 convolution filters in the first layer followed by a max pooling layer with a pool size of 3 × 3 and 30 convolution filters in the second convolution layer with another max pooling layer having a size of 2 × 2 following it. The framework of the CNN network is shown in Fig. 4. The convolution operator determines the output neurons that have to be connected to the input region; this happens through dot product calculation between the scalars, the weight matrix, and a specific small region for linking in the input volume. Every pooling layer reduces the feature filter dimensions over the neighboring pixels of the feature filters and hence increases the ability to remain steady even under instability on the inputs. The rate of decreasing the dimensions of the features filters with the pooling strides depends on the function [28].

$$y = f(x) = f_L(\ldots, f_2(f_1(x, u_1), u_2, \ldots), u_L) \tag{3}$$

In each of the layers in the architecture, to learn from the parameters, both forward and backward propagation techniques are computed, and several parameters are put to accomplish this. Each function takes data as a form of input matrix of the size of M × N and a weight parameter vector; they produce an output data. The first input tells a CT scan image needs to undergo processing while the rest are in-between feature filters. For every convolution layer the initial input filter of the weight matrix is randomly initiated but with predefined initialized filter size. The output of the convolution operation with this initial bank of filters, y, is [29, 30]:

$$y_{i'j'} = sum_{ijk}\left(u_{(ijk')} x_{i+i', j+j'}\right) \tag{4}$$

The size of the feature filter can be calculated from the following formula [28]:

$$x_{i+1}^{size} = \left(\frac{x_i^{size} - F_i + 2*Padding\_size}{Stride\_length} + 1\right) \tag{5}$$

Along with this, each pixel value of the feature filter maps is kept as a parameter to a nonlinear activating process. Rectified linear unit (ReLU) activation function is applied to threshold the data with zero:

$$y_{ij} = max(0, x_{ij}) \quad (6)$$

The activation operation alters the expanse of the feature filter maps. Moreover, to decrease the expanse of the feature filter map, maximum pooling layer is applied to integrate nearby neighboring feature values into decreased size samples and decrease the effect of noise. Commonly max pooling is used to select the most significant component in the neighborhood of the pixels. The pooling stride controls the rate of decreasing of pixels:

$$y_{ij} = max\{y_{i'j'} : i \leq i' + p, j \leq j' < j + q\} \quad (7)$$

There is a dropout operator to work with bias and variance in CNN networks. In this, randomly chosen neurons and their connections are dropped out from the neural network framework during the training step.

Once every layer passes the forward propagation, the backward step proceeds to tune the parameters of the feature maps like the weight matrix $u = (u1,...,uL)$ in such a way as to get the function $Q = f(x, u)$ sustain the minimum loss, $loss(Q, \hat{Q})$ where $Q = (Q1, ..., Q_i, ...)$ is the output value of the input.

The loss function is

$$Loss(u) = \frac{1}{n}\sum_{i=1}^{n} loss(Q_i, f(x_i, u)) \quad (8)$$

To minimize this loss function, we use an approach called gradient descent that quantifies the gradient of loss function at a certain weight and then updates the weights based on a learning parameter, accordingly along the direction of fast descent through finding the minima of the loss function:

$$u^{t+1} = u^t - \alpha \frac{\partial f}{\partial w}(u^t) \quad (9)$$

The slices of the images are applied to train the CNN model, and the classification of the CT images is performed with a classifier using the softmax activation function, which calculated normalized to be predicted class vector probabilities defined by the softmax function termed as the predicted output [31, 32]:

$$f_{softmax}(Q) = \frac{e^{Q_j}}{\sum_k e^{Q_k}} \quad (10)$$

The function converts the vector of scores from Q and compresses it to a vector between zero and one such that the sum of the values is equal to one. The class scores f is calculated through cross entropy loss values that are formulated using the equation [33, 34]:

$$Loss_i = -f_{y_i} + \log \sum_j e^{f_j} \qquad (11)$$

Here, f_j refers to the jth element of the vector outputs.

7 Results and discussion

The convolutional neural network can be learned from the given test images formed by splitting the hundred image into training and test images, to classify whether the images are rightly classified when it comes to age and contrast material used if there are any misclassifications done. The model was seen initially to show overfitting, which had been overcome later by using dropout layers and reducing the number of epochs for training. The output was coming in a multiclass form, which needed to be transformed into a single class output for analysis of the result. The accuracy achieved during the classification amounted to 89.99%. During the training stage of the convolutional neural network structure, in each epoch, the categorical cross-entropy loss function reduces due to the backpropagation during the training stage. The loss values for each epoch value are shown in the following graph [35].

From Fig. 5, it can be seen that as the number of epochs increases, the cross-entropy loss keeps on decreasing due to the gradient descent algorithm. Due to the forward propagation steps and corresponding backpropagation steps, the model learned from the weights given, and the corresponding accuracy of the

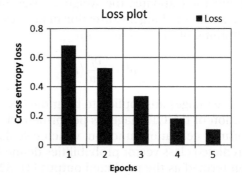

Fig. 5 The plot shows how cross-entropy loss changes with the number of epochs being trained by the network.

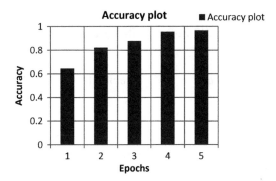

Fig. 6 The plot shows how the accuracy of the learning algorithm changes with the number of epochs trained.

model kept on increasing. The accuracy values for each epoch are shown in the following graph [25].

From Fig. 6, it can be perceived that the accuracy of the model is increasing as the number of epochs increases. The average precision, average recall, and average F1 score from the classification model are tabulated in Table 2.

The error matrix for the classification is given in Table 3. The classification that was applied to the 10 test images from the dataset had 9 such images correctly classified regarding age and contrast used. One such image was wrongly classified regarding contrast used.

Table 2 Average precision, recall, and F1 score values.

Average precision	Average recall	Average F1 score
0.925	0.90	0.903297

Table 3 Confusion matrix.

	Classified true	Classified false
True	6	1
False	0	3

Table 4 AUC-ROC, specificity, and sensitivity analysis.

AUC-ROC score	Specificity	Sensitivity
0.92857	80%	1%

The receiver operating characteristics curve (ROC) was drawn for the following test images, and the area under the ROC curve (AUC) value was derived from it. Along with the result, from the confusion matrix, true-positive rates and false-positive rates were calculated, and with that the specificity and sensitivity percentages were derived, and their results are tabulated in Table 4 [36, 37].

During training, it was seen that overfitting was seen for a number of epochs greater than six. It resulted in correct classification for all the test cases, evident from having a training accuracy of one. Furthermore, to rectify the errors, dropout layers were applied; it resulted in a perfectly fitted classified data [38].

8 Conclusion

Computed tomography (CT) images can be very risky when it comes to the application of contrast material. The risk of radiation can increase to some extent when it comes to contrast. The CT scan images also show varying trends for patients of a different age. The high accuracy of the model provided by our proposed method can give classification results for new unseen test images when it comes to age and contrast application that can work wonders in helping doctors looking for trends in the CT images. The doctors can use the classified images to tell the patient about their risk toward specific radiation if any that can be caused due to the procedure.

References

[1] M. Nakrani, G. Sable, U.B. Shinde (2019). Detection of lung nodules in computed tomography image using deep machine learning: a review. In: *Proceedings of International Conference on Communication and Information Processing (ICCIP)*.

[2] ACR, R, Patient Safety—Contrast Material, [online] Radiologyinfo.org. Available at: https://www.radiologyinfo.org/en/info.cfm?pg=safety-contrast, 2018.

[3] J.E. Barnes, Characteristics and control of contrast in CT, Radiographics 12 (4) (1992) 825–837.
[4] M.A. Mazurowski, M. Buda, A. Saha, M.R. Bashir, Deep learning in radiology: an overview of the concepts and a survey of the state of the art, arXiv preprint arXiv:1802.08717 (2018).
[5] D.J. Brenner, Estimating cancer risks from pediatric CT: going from the qualitative to the quantitative, Pediatr. Radiol. 32 (4) (2002) 228–231.
[6] T. Szabó, P. Barsi, P. Szolgay, Application of analogic CNN algorithms in telemedical neuroradiology, in: Cellular Neural Networks and Their Applications, 2002, pp. 579–586.
[7] M.C. Chamberlain, J.A. Murovic, V.A. Levin, Absence of contrast enhancement on CT brain scans of patients with supratentorial malignant gliomas, Neurology 38 (9) (1988) 1371.
[8] A. Lisowska, E. Beveridge, K. Muir, I. Poole, Thrombus detection in CT brain scans using a convolutional neural network, in: Bioimaging, 2017, pp. 24–33.
[9] M. Havaei, A. Davy, D. Warde-Farley, A. Biard, A. Courville, Y. Bengio, C. Pal, P. M. Jodoin, H. Larochelle, Brain tumor segmentation with deep neural networks, Med. Image Anal. 35 (2017) 18–31.
[10] R. Biswas, A. Vasan, S.S. Roy, Dilated deep neural network for segmentation of retinal blood vessels in fundus images, Iran. J. Sci. Technol., Trans. Electr. Eng. (2019) 1–14.
[11] V.E. Balas, S.S. Roy, D. Sharma, P. Samui (Eds.), Handbook of Deep Learning Applications, In: vol. 136, Springer, 2019.
[12] S.S. Roy, R. Sikaria, A. Susan, A deep learning based CNN approach on MRI for Alzheimer's disease detection, Intelligent Decision Technologies 13 (4) (2019) 495–505.
[13] A. Bose, S.S. Roy, V.E. Balas, P. Samui, Deep learning for brain computer interfaces, in: Handbook of Deep Learning Applications, Springer, Cham, 2019, pp. 333–344.
[14] E.I. Zacharaki, S. Wang, S. Chawla, D. Soo Yoo, R. Wolf, E.R. Melhem, C. Davatzikos, Classification of brain tumor type and grade using MRI texture and shape in a machine learning scheme, Magn. Reson. Med. 62 (6) (2009) 1609–1618.
[15] M.C. Chamberlain, A.D. Sandy, G.A. Press, Leptomeningeal metastasis: a comparison of gadolinium-enhanced MR and contrast-enhanced CT of the brain, Neurology 40 (3 part 1) (1990) 435.
[16] W. Zhang, R. Li, H. Deng, L. Wang, W. Lin, S. Ji, D. Shen, Deep convolutional neural networks for multi-modality isointense infant brain image segmentation, NeuroImage 108 (2015) 214–224.
[17] J. Kleesiek, G. Urban, A. Hubert, D. Schwarz, K. Maier-Hein, M. Bendszus, A. Biller, Deep MRI brain extraction: a 3D convolutional neural network for skull stripping, NeuroImage 129 (2016) 460–469.
[18] M. Simonovsky, B. Gutiérrez-Becker, D. Mateus, N. Navab, N. Komodakis, A deep metric for multimodal registration, in: International Conference on Medical Image Computing and Computer-Assisted Intervention, Springer, Cham, 2016, October, pp. 10–18.
[19] P. Wu, Z. Cui, Z. Gan, F. Liu, Three-dimensional ResNeXt network using feature fusion and label smoothing for hyperspectral image classification, Sensors 20 (6) (2020) 1652.
[20] J. Li, W. Monroe, T. Shi, S. Jean, A. Ritter, D. Jurafsky, Adversarial learning for neural dialogue generation, arXiv preprint arXiv:1701.06547 (2017).
[21] H.I. Suk, S.W. Lee, D. Shen, Alzheimer's Disease Neuroimaging Initiative, Hierarchical feature representation and multimodal fusion with deep learning for AD/MCI diagnosis, NeuroImage 101 (2014) 569–582.

[22] A. Payan, G. Montana, Predicting Alzheimer's disease: a neuroimaging study with 3D convolutional neural networks, arXiv preprint arXiv:1502.02506 (2015).
[23] S.M. Plis, D.R. Hjelm, R. Salakhutdinov, E.A. Allen, H.J. Bockholt, J.D. Long, H.J. Johnson, J.S. Paulsen, J.A. Turner, V.D. Calhoun, Deep learning for neuroimaging: a validation study, Front. Neurosci. 8 (2014) 229.
[24] T. Brosch, R. Tam, Alzheimer's Disease Neuroimaging Initiative, Manifold learning of brain MRIs by deep learning, in: International Conference on Medical Image Computing and Computer-Assisted Intervention, Springer, Berlin, Heidelberg, 2013, September, pp. 633–640.
[25] H. Huang, X. Hu, Y. Zhao, M. Makkie, Q. Dong, S. Zhao, L. Guo, T. Liu, Modeling task fMRI data via deep convolutional autoencoder, IEEE Trans. Med. Imaging 37 (7) (2017) 1551–1561.
[26] K. Clark, B. Vendt, K. Smith, J. Freymann, J. Kirby, P. Koppel, S. Moore, S. Phillips, D. Maffitt, M. Pringle, L. Tarbox, F. Prior, The cancer imaging archive (TCIA): maintaining and operating a public information repository, J. Digit. Imaging 26 (6) (December, 2013) 1045–1057.
[27] B. Albertina, M. Watson, C. Holback, R. Jarosz, S. Kirk, Y. Lee, J. Lemmerman, Radiology Data from The Cancer Genome Atlas Lung Adenocarcinoma [TCGA-LUAD] Collection. The Cancer Imaging Archive(2016). https://doi.org/10.7937/K9/TCIA.2016.JGNIHEP5.
[28] X.W. Gao, R. Hui, Z. Tian, Classification of CT brain images based on deep learning networks, Comput. Methods Prog. Biomed. 138 (2017) 49–56.
[29] H. Sugimori, Classification of computed tomography images in different slice positions using deep learning, J. Healthc. Eng. 2018 (2018).
[30] N.H.M. Duy, N.M. Duy, M.T.N. Truong, P.T. Bao, N.T. Binh, Accurate brain extraction using active shape model and convolutional neural networks, arXiv preprint arXiv:1802.01268 (2018).
[31] H. Greenspan, B. Van Ginneken, R.M. Summers, Guest editorial deep learning in medical imaging: overview and future promise of an exciting new technique, IEEE Trans. Med. Imaging 35 (5) (2016) 1153–1159.
[32] G. Litjens, T. Kooi, B.E. Bejnordi, A.A.A. Setio, F. Ciompi, M. Ghafoorian, J.A. van der Laak, B. Van Ginneken, C.I. Sánchez, A survey on deep learning in medical image analysis, Med. Image Anal. 42 (2017) 60–88.
[33] D. Nie, R. Trullo, J. Lian, C. Petitjean, S. Ruan, Q. Wang, D. Shen, Medical image synthesis with context-aware generative adversarial networks, in: International conference on medical image computing and computer-assisted intervention, Springer, Cham, 2017, September, pp. 417–425.
[34] P. Arena, A. Basile, M. Bucolo, L. Fortuna, Image processing for medical diagnosis using CNN, Nucl. Instrum. Methods Phys. Res., Sect. A 497 (1) (2003) 174–178.
[35] S. Vieira, W.H. Pinaya, A. Mechelli, Using deep learning to investigate the neuroimaging correlates of psychiatric and neurological disorders: methods and applications, Neurosci. Biobehav. Rev. 74 (2017) 58–75.
[36] A. Fieselmann, M. Kowarschik, A. Ganguly, J. Hornegger, R. Fahrig, Deconvolution-based CT and MR brain perfusion measurement: theoretical model revisited and practical implementation details, J. Biomed. Imaging 2011 (2011) 14.
[37] J.L. Claves, S.W. Wise, K.D. Hopper, D. Tully, T.R. Ten Have, J. Weaver, Evaluation of contrast densities in the diagnosis of carotid stenosis by CT angiography, AJR Am. J. Roentgenol. 169 (2) (1997) 569–573.
[38] P.Y. Lau, F.C. Voon, S. Ozawa, The detection and visualization of brain tumors on T2-weighted MRI images using multiparameter feature blocks, in: Engineering in Medicine and Biology Society, 2005. IEEE-EMBS 2005. 27th annual international conference of the IEEE, 2006, January, pp. 5104–5107.

Data analytics in IOT-based health care

Azadeh Zamanifar
Computer Engineering Department, University of Science and Culture, Tehran, Iran

1   Introduction

Internet of Things (IoT) health-care system has a high impact on mankind and society. It plays a significant role in the domain of pharmaceutical research and industry. The data involved are complex and heterogeneous, which makes them more challenging to understand and explore [1, 2].

ECG sensors are used for monitoring electrical impulses among heart muscle. Electromyogram (EMG) sensor is used for observing muscle activity. Electroencephalogram (EEG) sensor is used for recording electrical activity at the brain. Temperature sensor, core body temperature, and skin temperature are designed for health status prediction by measuring the body temperature. Breathing sensor is used for monitoring respiration [3]. IOT health care provides two-directional communications between a mobile node(s) and the remote server, which is not possible in traditional health-care systems. As the infrastructure of IOT health-care system is mobile IP-based wireless sensor network, power constraints are also one of the challenges of this environment [4]. Also, connected connectivity and low delay are the other significant challenges [5, 6]. The main goal in a IoT-based health-care system is to collect real-time data and take decision about the patient immediate help [7]. Thus the patient must be in the range of IoT network as a vital requirement. As mentioned the other important issue in health-care systems is continuous connectivity of the mobile node(s) to the network.

Thus the health data can be collected accurately. Therefore detecting and predicting the patient health status can be done in more accurate manner. Benefits of adopting IoT eHealth can be summed up as follows [8]:

(1) All-including: whether people use IoT eHealth for health, exercise, safety, or beauty reasons, it has a solution for everyone needs.
(2) Very smooth fusion with different technologies: IoT eHealth enables different technologies to work together perfectly.
(3) Ability to use big data collection and learning: This is done to efficiently analyzing the patient data.
(4) Easy to use: The patients do not need any complex guidance to use the system.
(5) Cost reduction: IoT health care system removes the need of patient to stay in a hospital. Thus the overall cost of continues monitoring of system reduces.
(6) Doctors can be more involved: Doctors receive online data, and they are notified by the condition sooner.
(7) Availability: The access to eHealth data/services is possible in anytime, anywhere without any inference.
(8) Online help: IoT eHealth enables 24/7 online access to health specialists such as doctors, skin doctors, and many other professionals.
(9) International working team effort: Health professionals around the world are connected via IoT eHealth community. This enables patients to have more access to international facilities at their fingertips (anywhere and anytime).

Biomedical knowledge has a high impact on mankind and society. It plays a significant role in the domain of pharmaceutical research and industry. The data involved are complex and heterogeneous, which makes them more challenging to understand and explore. In recent years, with an increase in the amount of biomedical data, there is rising opportunity for exploring and learning about their interactions and effects. Data preparation is always an enormous and time-consuming task in biomedical field. The process involves assembling of data from multiple reliable sources, analyzing their statistics, and processing them into some specific format, with satisfying the requirements of the associated biomedical problem.

In the following section, we classify the related works that have been done before. Based on the literature, we can divide the works to two categories as follows:
(1) off-line data analytic
(2) online data analytic

2 Off-line data analytic in IoT-based health care

Data analytic that are done in IoT-based health-care solution try to detect or predict the abnormal behavior of the patient. The online data that are collected from patient in health-care system can be further analyzed off-line. The biosensors that are used by patient are responsible for gathering health data. The other features that are important are age, sex, and other historical data about the health status of the patient. There are also another important data that are available by the help of IoT technology: the environmental status of the patient. These data are collected by the help of environmental sensors like movement sensors and sensors that are deployed in different part of the environment. These sensors can monitor the information that can help us to diagnose the activity of the patient. Hidden Markov model (HMM) [9] is one of the classic parts of the theory of Bayesian networks that is used [10] in various time series data, like speech, words, activity recognition, and gesture that are modeled. These are related to different patient recognition tasks [11]. HMM performs more accurately compared with artificial neural network (ANN) [12]. HSMM is an extension of HMM designed to take in to consideration the duration of staying in a state [13]. In HSMM, it may take more time to accurately determine the output [14]. Fuster-Parra et al. diagnose cardiovascular risk by using 23 features. They applied several classifiers. They found that by using Bayesian network, they got the best accuracy [15]. Sirvana et al. use different classifier like rule-based approaches, a decision tree, naive Bayes, and an artificial neural network on large data set of health care consisting of features related to heart attack. The results showed that naive Bayes approach is better than the others.

In Srinivas et al. [16], they construct a model for predicting online health status per user, based on the temporal and spatial information of the patient.

These solutions can better use the IoT characteristics to do recognize an abnormal status [17]. Furthermore, they cannot predict (and can only detect) an abnormal status. Most of these methods update their models very rarely. Typical IoT health care studies can only detect the health status of the patient by using activity recognition methods. The precision of the health status prediction is low in these studies. Some other studies [18, 19] have focused on collecting in door data from sensors and using SVM classifier to categorize abnormal behavior. However, they manually tag the data set. Moreno et al. [17] proposed an abnormal situation detection in nursing home.

Yin et al. [20] propose a bidirectional human-machine interface. It includes an extended physiological proprioception (EPP) system and a neurofuzzy controller. It is constructed based on imitating the biological closed-loop control system of the human body. The exoskeleton is an electromechanical device worn by patient. It is designed according to shape and function of the human body. It helps human limb and/or muscles, joints, or skeletal parts that are weak, ineffective, or injured because of a disease or a neurological condition.

Kumara et al. [21] introduce a high-level model in IoT-based health-care system that collected data of mobile patient, which are transmitted to a social system. The social system groups similar patients based on their behavior. This is done for symptom analysis of patients. The model can predict anomalies. But it is not implemented. The literature only introduces a high-level model.

Most of the related works are based on finding behavioral patterns unfamiliar to normal behavior, thus the abnormality can be detected [22, 23]. Many studies discuss about using learning methods to detect critical events like falling [24–29]. Clustering algorithms have also been used to identify abnormal behavior patterns [30, 31].

Xu et al. [32] proposed a solution for decentralized storage and analysis of personal health data by using multilevel cloud-based architecture. Health data are collected from smart devices and transferred to the cloud side. They introduce process mining and similar patient searching algorithms to support personalized treatment plan selection. A similarity calculation formula is designed to choose similar historic treatment plans, and then the similar plans are recommended to doctors for references in clinical decision-making. However, our goal is quite different and is focused on predicting health status of the patient based on ECG sensors and the history of patient's movement. We do not suggest any treatment plan too.

Mobile ECG [33] system uses smart mobile phones as the base station for ECG measurement and analysis. Smart mobile phone receives ECG data from mobile ECG recording devices via Bluetooth. Mobile phones store and analyze the ECG data and forward them to the medical professionals if necessary. The graphical interface is capable of displaying ECG recordings and heart rate. They do not report the accuracy of their result too.

Yang et al. [34] propose a series of validation rules with uncertainty threshold parameters and reliability indicators to validate the life-logging data in an IoT-enabled health-care system. They reach 75% accuracy in their result. The experiments show that they filter near 75% of irregular data. However, our paper predicts the health status of the patient. We can further use the mentioned technique to gain a better result.

The University of Virginia's (UVA's) ALARM-NET system, described in [35], is based on the use of a multitiered heterogeneous sensor network designed to monitor people. The application of ALARM-NET is that of having an automatic real-time monitoring system, deployed in an assisted-living facility or residence, to report on the residents' health, activity, and environment. The data from the body sensors are broadcast and single hop to the nearest stationary sensor motes in the second tier of the ALARM-NET system. The second tier of motes uses multihop communication based on the standard TinyOS 1.1.15 configuration to send and forward data from the body sensor network to a node in the IP-based network. The information collected from the patient is routed to the nearest person/node. However, the applicability and the goal of the paper are quite different from ours.

Suryadevara et al. [36, 37] monitor the activity of older adults to determine normal/abnormal health status. They define wellness function in elderly patient by means of real-time activity behavior recognition. Six different sensors are deployed in environment, and the unnormal condition can be detected. However, they achieve low accuracy.

Dohr et al. [38] developed MobiCare. The structure of Mobicare consists of different components and the interaction between them. The components of the system include (1) every self-organizing service that is included in sensors and devices and (2) services for automatic update and remote configuration. The proposed system cannot predict the health status.

Gayathri et al. [39] detect abnormal health condition. They deploy hierarchical Markovian logic network. The assumption that is taken is that there are several different sensors deployed within the house. The goal of the study is obtaining the abnormal condition: (1) sensor's data, (2) the time of patient arrival, (3) the duration that a patient stays in a room, and (4) the possibility that patient does more than one activity at the same time. They deploy two learning approaches. The first method is used for activity detection and classification. The second method can extract the relevancy between different features. Thus the anomaly in health status can be recognized. This solution has noticeable overhead.

Most of the solutions in IoT health-care systems [40] needs that patients wear different body sensors, which are used to detect the activity of patient.

Many learning methods [41] can be applied in different part of patient health status detection and prediction. Applying many different learning methods in health-care system can help us in recognizing normal activity and knowing the relationship between different feature of the environment and the normal behavior of the patient. Thus unnormal pattern can be recognized. However, offline methods cannot predict the abnormal health status of patient or elderly people.

3 Online data analytic in IoT-based health-care systems

Hossain et al. [42] have used electrocardiogram (ECG) sensor for monitoring health status of the patient. This is done due to the fact that measuring ECG is an important factor for body. By monitoring ECG signals continuously, a healthcare expert can detect potential risk. ECG signals are recorded via portable ECG recording devices at home or outdoors. ECG signal can be transmitted to smart phones or desktops via any communication technology. On the other side an application rejects or filters any noise from the received signal and provides a secure mechanism for authentication and authorization. Heartbeat can also be extracted from ECG signal. The other features like temporal and spatial properties are considered, and they are used for classification approach like support vector machine. The remote health-care system receives these, and the expert takes a decision and sends the decision by notification to the patient [43].

Meng et al. propose an online model for daily habitant and anomaly detection (ODHMAD). Their architecture consists of different components: one component for activity recognition and the other one for habit modeling [44]. The model does online processing of received data. But this work cannot predict anomalies.

Yassine et al. [8] combine IoT and fog technology and analyze big data of smart home. Fog computing provides huge and fast enough computing environment. The system gets data from various IoT systems and stores them in cloud. Analyzing them provides a good prediction model of anomaly.

The previous work of the author [45] introduced an approach for online prediction of health status of patient. The environment is divided into equal sized cell. At the center of each cell, a static node is positioned. The static node is a sensor that sends the time of arrival of the patient, the duration of staying within a cell, and the ECG data of the patient to the gateway. The ECG data are sent periodically. This work is done for every patient within the indoor environment. For 1 month the data are collected. Then, at the gateway, the HSMM learning method is used to train the model. HSMM is a hidden semi-Markov model [46], which takes in to consideration the duration of staying within a cell. After constructing the model, it splits around the area, and each static node at the center of each cell receives a portion of the model that is related to itself. The communication between static nodes is done with the special routing path. A tree is constructed with all static nodes, in which the static nodes at the center of each cell are leaves of the tree. A mobile node will communicate with the gateway via leaf nodes. Since people in elderly households have almost predefined daily routine, they recognize activities in each cell implicitly without using any biological and environmental sensors to record the patient's physical state, objects, or environmental conditions. The proposed method constructs a distributed online prediction of mobile node data in IoT health-care applications. Existing approaches that detect a patient's health status using activity recognition methods are not cost-effective, due to their need for using lots of body. The results show the effectiveness, time of arrival within a cell, ECG signal, duration, and location in increasing the accuracy of our proposed method.

Manogaran [47] proposes an architecture for big sensor data storing and computation in health-care system. They use a map reduce architecture to efficiently process the huge data. They use scalable logistic regression to process large data in distributed manner. They deploy Apache Mahout with Hadoop Distributed File system to do so [47]. The solution helps for no stop monitoring the health status of the patient. It consists of different stages from (1) data gathering and transmitting. The collected data consist of heart rate, blood pressure and sugar, and body temperature. Amazon S3 bucket is used to store the data in a continuous manner. Then, Amazon EMR with HBase database is deployed to store them permanently. The data are categorized according to their importance. By using the stochastic gradient algorithm with regression, the prediction model for heart and diabetic patient is constructed. The profile info of the patients is also extracted from Cleveland Hearth Disease DB for training.

Yuanjo [48] proposes an integrated framework for prognostic and management of different health conditions. It is based on IoT and convolutional neural network (CNN). The framework proposes a guidance for manufacturers in three different levels (strategic, tactical, and operational). The status of IoT devices is monitored for determining health index. At the operational level the network gathers the data based on antireference from the group of devices.

4 Open research and challenges

Real-time data analysis in IoT-based health-care system has many challenges. We briefly introduce them in this section. IoT-based health-care system needs real-time analysis to detect or predict the health status of the patient more accurately. Real-time analysis is in contrast with machine learning methods that are heavy in nature, and they are not cost effective in terms of time. Real-time analysis of these data can improve machine health and lead to defect-free product manufacturing [46, 49]. The one concern of IoT health-care system is the continuous collection of data from the sensors and analyzing at the same time. It is a challenging task as the limitation that exists in both storage and communication media. As every patient has its own characteristic and health data, in an environment with n patients, we have to consider enormous data storage, computation, and communication capacity. The other concern is maintaining the health status model update. It raises a question that what is the best time to update a model. As it requires time and cost to do so, finding an optimized time is a necessity.

5 Conclusion

The health-care industry has created large amount of data created from record keeping and patient-related data. In today's digital world, it is required that these data should be put into a computer. To improve the quality of health care by the costs, it's necessary that large amount of data created should be analyzed effectively to answer new challenges. In almost the same way, government also creates quadrillion bytes of data every day. It needs a technology that helps to do a real-time analysis on the huge data set. This will help the government to provide value-added services to the people. Big data helps in discovering valuable decisions by understanding the data patterns and the relationship between them with the help of machine learning sets of computer instructions. The IoT paradigm has become very popular in recent years due to its numerous benefits [38, 50–53], including the omission of the human role in aggregating and analyzing data. IoT changes the many aspect of life aiming to delete the involvement of the operator or expert in taking a decision. In safety critical applications like health care, it needs that the task of the physician or nurse is accurately replaced by system. Thus we need more accurate decision and less delay IoT health-care system. Learning-based system produces some false-positive results, which is not the goal of the system. It seems that deploying more accurate learning-based methods could reach us to our destination in IoT-based health-care system, which is a more accurate and responsive system.

References

[1] V. Gazis, K. Sasloglou, N. Frangiadakis, P. Kikiras, Wireless sensor networking, automation technologies and machine to machine developments on the path to the internet of things, in: Informatics (PCI), 2012 16th Panhellenic Conference on, 2012, pp. 276–282.

[2] M.S. Shahamabadi, B.B.M. Ali, P. Varahram, A.J. Jara, A network mobility solution based on 6LoWPAN hospital wireless sensor network (NEMO-HWSN), in: Proceedings of the 2013 Seventh International Conference on Innovative Mobile and Internet Services in Ubiquitous Computing, IMIS '13, IEEE Computer Society, Washington, DC, 2013, pp. 433–438.

[3] P. Kulkarni, Y. Oztürk, Requirements and design spaces of mobile medical care, ACM SIGMOBILE Mob. Comput. Commun. Rev. 11 (3) (2007) 12–30.

[4] A. Zamanifar, E. Nazemi, M. Vahidi-Asl, DMP-IOT: a distributed movement prediction scheme for iot health-care applications, Comput. Electr. Eng. 58 (2017) 310–326.

[5] A. Zamanifar, E. Nazemi, An approach for predicting health status in IOT health care, J. Netw. Comput. Appl. 134 (2019) 100–113.

[6] A. Zamanifar, E. Nazemi, EECASC: an energy efficient communication approach in smart cities, Wirel. Netw. (2018) 1–16.

[7] H. Alemdar, C. Ersoy, Wireless sensor networks for healthcare: a survey, Comput. Netw. 54 (15) (2010) 2688–2710.

[8] A. Yassine, S. Singh, M.S. Hossain, G. Muhammad, IOT big data analytics for smart homes with fog and cloud computing, Future Gener. Comput. Syst. 91 (2019) 563–573.

[9] S. Akoush, A. Sameh, Mobile user movement prediction using bayesian learning for neural networks. in: Proceedings of the 2007 International Conference on Wireless Communications and Mobile Computing, IWCMC '07, 2007, pp. 191–196, https://doi.org/10.1145/1280940.1280982.

[10] J. Whittaker, Graphical Models in Applied Multivariate Statistics, Wiley Publishing, 2009.

[11] L. Gellert, L. Vintan, Person movement prediction using hidden Markov models, Stud. Inform. Control 15 (2006) 17–30.

[12] Y. Lee, L. Ow, D. Ling, Hidden Markov models for forex trends prediction. in: Information Science and Applications (ICISA), 2014 International Conference on, 2014, pp. 1–4, https://doi.org/10.1109/ICISA.2014.6847408.

[13] J.D. Ferguson, Variable duration models for speech, in: Proceedings of the Symposium on the Application of HMMs to Text and Speech, 1980, pp. 143–179.

[14] S.-Z. Yu, H. Kobayashi, A hidden semi-Markov model with missing data and multiple observation sequences for mobility tracking, Signal Process. 83 (2) (2003) 235–250.

[15] P. Fuster-Parra, P. Tauler, M. Bennasar-Veny, A. Ligeza, A. Lopez-Gonzalez, A. Aguilo, Bayesian network modeling: a case study of an epidemiologic system analysis of cardiovascular risk. Comput. Methods Prog. Biomed. 126 (2016) 128–142, https://doi.org/10.1016/j.cmpb.2015.12.010.

[16] K. Srinivas, B.K. Rani, A. Govrdhan, Applications of data mining techniques in healthcare and prediction of heart attacks, Int. J. Comput. Sci. Eng. 2 (02) (2010) 250–255.

[17] L. Moreno-Fernandez-de, J.M. Lopez-Guede, M. Graña, J.C. Cantera, Real Prediction of Elder People Abnormal Situations at Home, Springer International Publishing, Cham, 2017, pp. 31–40.

[18] H. Gottfried, K. Aghajan, B.-Y. Wong, J.C. Augusto, H.W. Guesgen, T. Kirste, M. Lawo, Spatial health systems, in: Smart Health, Springer, 2015, pp. 41–69.

[19] V.R. Jakkula, D.J. Cook, Detecting anomalous sensor events in smart home data for enhancing the living experience, Artif. Intell. Smarter Living 11 (201) (2011) 1.

[20] Y.H. Yin, Y.J. Fan, L.D. Xu, EMG and EPP-integrated human-machine interface between the paralyzed and rehabilitation exoskeleton, IEEE Trans. Inf. Technol. Biomed. 16 (4) (2012) 542–549.

[21] S. Kumara, L. Cui, J. Zhang, Sensors, networks and internet of things: research challenges in health care. in: Proceedings of the 8th International Workshop on Information Integration on the Web: In Conjunction with WWW 2011, IIWeb '11, ACM, New York, NY, USA, 2011, pp. 2.1–2.4, https://doi.org/10.1145/1982624.1982626.

[22] S.S. Khan, M.E. Karg, J. Hoey, D. Kulic, Towards the detection of unusual temporal events during activities using HMMS, in: Proceedings of the 2012 ACM Conference on Ubiquitous Computing, ACM, 2012, pp. 1075–1084.

[23] F.J. Ordóñez, P. de Toledo, A. Sanchis, Sensor-based bayesian detection of anomalous living patterns in a home setting, Pers. Ubiquit. Comput. 19 (2) (2015) 259–270.

[24] J. Yin, Q. Yang, J.J. Pan, Sensor-based abnormal human-activity detection, IEEE Trans. Knowl. Data Eng. 20 (8) (2008) 1082–1090.

[25] J. Cheng, X. Chen, M. Shen, A framework for daily activity monitoring and fall detection based on surface electromyography and accelerometer signals, IEEE J. Biomed. Health Inform. 17 (1) (2013) 38–45.

[26] B. Mirmahboub, S. Samavi, N. Karimi, S. Shirani, Automatic monocular system for human fall detection based on variations in silhouette area, IEEE Trans. Biomed. Eng. 60 (2) (2013) 427–436.

[27] S. Rakhecha, K. Hsu, Reliable and secure body fall detection algorithm in a wireless mesh network, in: Proceedings of the 8th International Conference on Body Area Networks, ICST (Institute for Computer Sciences, Social-Informatics and Telecommunications Engineering), 2013, pp. 420–426.

[28] J. Chen, K. Kwong, D. Chang, J. Luk, R. Bajcsy, Wearable sensors for reliable fall detection, in: Engineering in Medicine and Biology Society, 2005. IEEE-EMBS 2005. 27th Annual International Conference of the, IEEE, 2006, pp. 3551–3554.

[29] Q. Li, J.A. Stankovic, M.A. Hanson, A.T. Barth, J. Lach, G. Zhou, Accurate, fast fall detection using gyroscopes and accelerometer-derived posture information, in: Wearable and Implantable Body Sensor Networks, 2009. BSN 2009. Sixth International Workshop on, IEEE, 2009, pp. 138–143.

[30] M.-S. Lee, J.-G. Lim, K.-R. Park, D.-S. Kwon, Unsupervised clustering for abnormality detection based on the tri-axial accelerometer, ICCAS-SICE, 2009, IEEE (2009) 134–137.

[31] A. Lotfi, C. Langensiepen, S.M. Mahmoud, M.J. Akhlaghinia, Smart homes for the elderly dementia sufferers: identification and prediction of abnormal behaviour, J. Ambient Intell. Humaniz. Comput. 3 (3) (2012) 205–218.

[32] B. Xu, L. Xu, H. Cai, L. Jiang, Y. Luo, L. Gu, The design of an m-health monitoring system based on a cloud computing platform, Enterp. Inf. Syst. 11 (1) (2017) 17–36.

[33] H. Kailanto, E. Hyvarinen, J. Hyttinen, Mobile ECG measurement and analysis system using mobile phone as the base station, in: Pervasive Computing

Technologies for Healthcare, 2008, Pervasive Health 2008. Second International Conference on, IEEE, 2008, pp. 12–14.

[34] P. Yang, D. Stankevicius, V. Marozas, Z. Deng, E. Liu, A. Lukosevicius, F. Dong, L. Xu, G. Min, Lifelogging data validation model for internet of things enabled personalized healthcare, IEEE Trans. Syst. Man Cybernet. Syst. 48 (1) (2018) 50–64.

[35] D. Wood, J.A. Stankovic, G. Virone, L. Selavo, Z. He, Q. Cao, T. Doan, Y. Wu, L. Fang, R. Stoleru, Context-aware wireless sensor networks for assisted living and residential monitoring, IEEE Netw. 22 (4) (2008).

[36] N.K. Suryadevara, S.C. Mukhopadhyay, Wireless sensor network based home monitoring system for wellness determination of elderly, IEEE Sensors J. 12 (6) (2012) 1965–1972.

[37] N.K. Suryadevara, S.C. Mukhopadhyay, R. Wang, R. Rayudu, Forecasting the behavior of an elderly using wireless sensors data in a smart home, Eng. Appl. Artif. Intell. 26 (10) (2013) 2641–2652.

[38] A. Dohr, R. Modre-Opsrian, M. Drobics, D. Hayn, G. Schreier, The internet of things for ambient assisted living. in: Information Technology: New Generations (ITNG), 2010 Seventh International Conference on, 2010, pp. 804–809, https://doi.org/10.1109/ITNG.2010.104.

[39] K. Gayathri, S. Elias, B. Ravindran, Hierarchical activity recognition for dementia care using Markov logic network. Pers. Ubiquit. Comput. 19 (2) (2015) 271–285, https://doi.org/10.1007/s00779-014-0827-7.

[40] S. Shaji, M.V. Ramesh, V.N. Menon, Real-time processing and analysis for activity classification to enhance wearable wireless ECG, in: Proceedings of the Second International Conference on Computer and Communication Technologies, Springer, 2016, pp. 21–35.

[41] P. Samui, S.S. Roy, V.E. Balas (Eds.), Handbook of Neural Computation, Academic Press, 2017.

[42] M.S. Hossain, G. Muhammad, Cloud-assisted industrial internet of things (IIOT)–enabled framework for health monitoring, Comput. Netw. 101 (2016) 192–202.

[43] A. Zamanifar, E. Nazemi, M. Vahidi-Asl, A mobility solution for hazardous areas based on 6LoWPAN, Mob. Netw. Appl. 23 (6) (2018) 1539–1554.

[44] L. Meng, C. Miao, C. Leung, Towards online and personalized daily activity recognition, habit modeling, and anomaly detection for the solitary elderly through unobtrusive sensing, Multimed. Tools Appl. 76 (8) (2017) 10779–10799.

[45] A. Zamanifar, E. Nazemi, M. Vahidi-Asl, A mobility solution for hazardous areas based on 6LoWPAN, Mob. Netw. Appl. (2017) 1–16.

[46] X.D. Huang, Y. Ariki, M.A. Jack, Hidden Markov Models for Speech Recognition, vol. 2004, Edinburgh University Press, Edinburgh, 1990.

[47] G. Manogaran, D. Lopez, C. Thota, K.M. Abbas, S. Pyne, R. Sundarasekar, Big data analytics in healthcare internet of things, in: Innovative Healthcare Systems for the 21st Century 2017, 2017, pp. 263–284.

[48] Y. Qu, et al., An integrative framework for online prognostic and health management using internet of things and convolutional neural network, Sensors 19 (10) (2019) 2338.

[49] G. Manogaran, et al., A new architecture of internet of things and big data ecosystem for secured smart healthcare monitoring and alerting system, Futur. Gener. Comput. Syst. 82 (2018) 375–387.

[50] I. Akkaya, P. Derler, S. Emoto, E.A. Lee, Systems engineering for industrial cyber-physical systems using aspects, Proc. IEEE 104 (5) (2016) 997–1012.

[51] P. Georgakopoulos, P. Jayaraman, Internet of things: from internet scale sensing to smart services, Computing 98 (10) (2016) 1041–1058.
[52] L.D. Xu, W. He, S. Li, Internet of things in industries: a survey. IEEE Trans. Ind. Inform. 10 (4) (2014) 2233–2243, https://doi.org/10.1109/TII.2014.2300753.
[53] L. Atzori, A. Iera, G. Morabito, The internet of things: a survey, Comput. Netw. 54 (15) (2010) 2787–2805.

Application of PCA based unsupervised FE to neurodegenerative diseases

Y.-H. Taguchi[a] and Hsiuying Wang[b]
[a]Department of Physics, Chuo University, Tokyo, Japan. [b]Institute of Statistics, National Chiao Tung University, Hsinchu, Taiwan

1 Introduction

Data science is a new tool to investigate genomic science. At the start, microarrays with individual probes were used to measure the amount of transcripts transcribed in the genome. Following this technology, so-called next generating sequencing (NGS) appeared. In contrast to microarrays specific to RNA measurements, NGS can measure DNA as well as RNA. Today, NGS can measure several kinds of epigenomes (i.e., modifications in the genome that can affect transcript without changing nucleotide sequences). Integrated analysis of several kinds of epigenomes together with RNA and DNA is known as multiomics.

The common feature of these measurements from the data science point of view is that they are "large p small n" problems, which means that there are only a small number of samples having many variables. Recently, machine learning techniques were introduced to data science that aim to analyze large amounts of data; the most popular strategy in machine learning is supervised learning. In supervised learning, functions are tuned to predict targets (e.g., classification labels) from provided data. Thus, we need a huge amount of samples to train functions to avoid overfitting. Overfitting means that functions are tuned to be too specific to a given data set to predict the target when using a newly obtained data set. This means that without a huge amount of samples, we cannot tune functions properly.

Unfortunately, a huge amount of samples cannot easily be obtained for genomic science due to many reasons. First, measurements of the genome are very expensive. Usually, the number of samples should be much larger than the number of variables in data science; in genomic science, the number of variables is that of genes (i.e., a few tens of thousands). Because of the expense of genomic measurements, it is unrealistic to gather as many samples as genes. Second, it is hard to collect as many samples as genes even when disregarding cost. For example, suppose that we would like to study a specific disease. It is ideal to have gene expression profiles from tissues where the disease takes place. Since it is usually necessary to perform surgery or needle biopsy, either of which might injure the human body, it is impossible to have gene expression profiles from tissues of healthy people, since there is no reason to injure healthy people only for collecting tissue gene expression profiles. Because of these difficulties, it is not very common to make use of supervised learning to train machine learning tools. Thus, it is better for us to have machine learning techniques that do not require a huge number of samples.

In order to invent a method fit to be applied to "large p small n" problems, we recently developed principal component analysis (PCA)-based unsupervised feature extraction (FE) [1]. In this method, multiomics data sets including gene expression profiles are treated with PCA and features (typically genes) are embedded into low-dimensional space with attributing PC score to features. Using attributed PC scores, features are selected based upon P-values obtained by assuming χ^2 distributions to PC scores. In the following, PCA-based unsupervised FE, in short, PCAUFE, will be applied to three independent data sets: (1) tissue mRNAs taken from amyotrophic lateral sclerosis (ALS) patients [2], (2) blood microRNA (miRNA) taken from ALS patients [3], and (3) tissue mRNA taken from Parkinson's disease (PD) patients [4]. Biological evaluations of selected genes using PCAUFE turned out to be reliable.

2 Materials and methods
2.1 PCAUFE

PCAUFE is an unsupervised method that can select genes based upon omics profile, typically gene expression profile, in a fully unsupervised manner. Suppose that $x_{ij} \in \mathbb{R}^{N \times M}$ corresponds to the ith feature (typically gene expression profile) of the jth sample (typically human patients or healthy controls). PCA is applied to x_{ij} such that PC score and PC loading are attributed to features and samples, respectively. In order that x_{ij} must be standardized as $\sum_i x_{ij} = 0$ and $\sum_i x_{ij}^2 = N$. Then, the lth PC score attributed to the ith feature is computed as the ith component of eigen vector, $\boldsymbol{u}_l \in \mathbb{R}^N$, of $N \times N$ matrix, $X^T X$, as

$$XX^T \boldsymbol{u}_l = \lambda_l \boldsymbol{u}_l$$

where X^T is a transposed matrix of X. Then, the lth PC loading attributed to the jth sample is computed from $\boldsymbol{u}_l$ as the jth component of $\boldsymbol{v}_l$ defined as

$$\boldsymbol{v}_l = X^T \boldsymbol{u}_l$$

$\boldsymbol{v}_l \in \mathbb{R}^M$ is also eigen vector of $X^T X \in \mathbb{R}^{M \times M}$ since

$$X^T X \boldsymbol{v}_l = X^T X X^T \boldsymbol{u}_l = X^T \lambda_l \boldsymbol{u}_l = \lambda_l \boldsymbol{v}_l$$

Using obtained $\boldsymbol{v}_l$ and $\boldsymbol{u}_l$, PCAUFE is performed as follows. First, among $\boldsymbol{v}_l$ attributed to samples, those associated with target properties (e.g., distinct values between patients and healthy controls) are selected. Then, P-values are attributed to the ith feature assuming χ^2 distribution to $\boldsymbol{u}_l$s as

$$P_i = \sum_l P_{\chi^2}\left[> \left(\frac{u_{li}}{\sigma_l}\right)^2 \right] \qquad (1)$$

where $P_{\chi^2}[>x]$ is cumulative χ^2 distribution whose argument is larger than x, σ_l is standard deviation of u_{li}, and summation of l is taken over only l associated with the selected $\boldsymbol{v}_l$ associated with target properties. P_is are corrected with *Benjamini* and *Hochberg* (BH) criterion [5] and features associated with corrected P-values less than 0.01 are selected.

2.2 Linear discriminant analysis using PC loading re-computed with only selected mRNAs or miRNAs

In this study, samples are discriminated with linear discriminant analysis (LDA) using PC loading, $\boldsymbol{v}'_l$, that are re-computed by applying PCA to only mRNAs or microRNAs selected by PCAUFE. PC loading, $\boldsymbol{v}'_l$, used for LDA are selected by applying categorical regression or t-test to obtained PC loadings. Leave-one-out cross validation (LOOCV) is used to evaluate performances.

2.3 Omics data sets

Omics profiles used in this study are as follows

2.3.1 Skeletal muscle mRNA expression profiles for ALS patients

mRNA expression profiles of skeletal muscle of 9 ALS patients and 10 healthy controls were downloaded from ArrayExpress [6] using the accession number *E*-MEXP-3260; all patients had sporadic ALS and presented with symptoms of limb onset. All samples were taken from human biopsies.

2.3.2 Serum miRNA expression profiles for ALS patients

miRNA expression profiles used in this study were obtained from gene expression omnibus (GEO) [7]. The file GSE52917_series_matrix.txt.gz included in "Series matrix" section associated with GEO ID GSE52917 was used. It consists of 9 familial ALS (fALS) patients analyzed with 6 arrays, 18 ALS mutant carriers without symptoms analyzed with 12 arrays, 18 sporadic ALS (sALS) patients and 17 healthy controls. In total 53 mRNA expression profiles were measured by microarray.

2.3.3 Substantia nigra mRNA expression profiles for PD patients

mRNA expression profiles are taken from GEO with three independent data sets associated with GEO ID, GSE20295, GSE20163, and GSE20164. With integrating three data sets, we can prepare those composed of 27 normal controls and 30 PD patients.

3 Results

3.1 Skeletal muscle mRNA expression profiles for ALS patients

After applying PCA to mRNA expression profiles, we noticed that v_2 is associated with significant distinction between 9 ALS patients and 10 healthy controls ($P = 9.44 \times 10^{-7}$ computed by *t*-test). Then we decided to select u_2 in order to attribute *P*-values to genes using Eq. (1). After correcting assigned *P*-values by BH criterion, 101 probes with GENBANK accessions were selected as those associated with adjusted *P*-values less than 0.01. Next, we tried to discriminate ALS patients from healthy controls using LDA with the second PC loading, v_2', re-computed by applying PCA to only these 101 genes selected (Table 1).

Table 1 Confusion matrix obtained by LDA applied to samples using the second PC loading re-computed by applying PCA to selected 101 probes.

Prediction	True ALS patients	Healthy controls
ALS patients	10	1
Healthy controls	0	8

The accuracy is as great as 0.95. The P-value computed by Fisher's exact test is 1×10^{-4} (Odds ratio is infinite); 101 genes selected can discriminate two groups successfully.

Next we try to validate biological significance of selected genes. Unfortunately, since the disease mechanism of ALS is unclear, evaluating selected genes biologically based upon the mechanism within them is not easy. Then we consider relationship with cancers in these genes; the association between ALS and cancer is often discussed [8]. Since we do not consider cancers at all when selecting 101 genes, if there is a relationship with cancers for these genes, it can be a side evidence that selected genes are biologically reasonable. In order to see this, after converting GENEBANK accessions to gene symbols, they are uploaded to OncoLnc [9] that can perform survival analysis based upon TCGA [10] data set.

Fig. 1 shows some examples of evaluating selected genes based upon cancer survival analyses performed by OncoLnc. It is obvious that some of genes selected for ALS are associated with cancer prognosis. As a result, we find 68 genes are associated with significant survival analysis of cancers (full list available as Table S4 [2]). In conclusion, we could successfully identify a set of genes that can be discriminated between healthy controls and ALS patients as well as those related to cancer survival probability. This can give us a genetic basis of observed association between ALS and various cancers.

3.2 Serum miRNA expression profiles for ALS patients

In the previous subsection, although we could successfully discriminate healthy controls from ALS patients, it made use of tissue mRNAs that are difficult to obtain. As such, it is better to

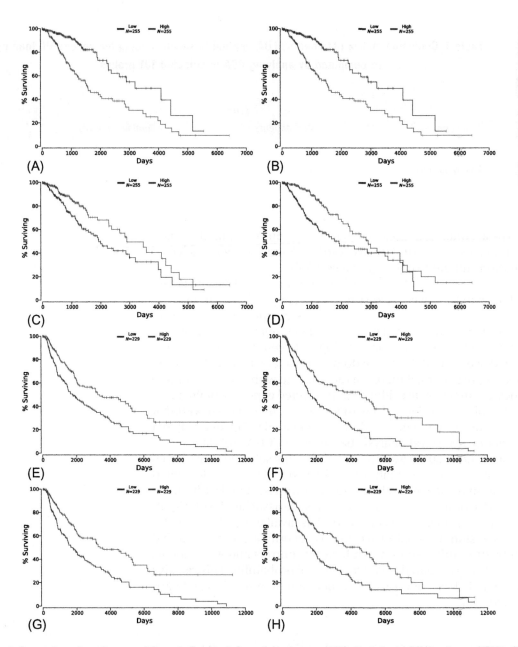

Fig. 1 Some examples of successful survival analysis for selected genes. LGG: Brain lower grade glioma, SKCM: Skin cutaneous melanoma. (A) LGG, RAD54L, $P = 1.17 \times 10^{-7}$, (B) LGG, ZNF443, $P = 1.78 \times 10^{-4}$, (C) LGG, dhps, $P = 3.12 \times 10^{-3}$, (D) LGG, RPL21, $P = 1.49 \times 10^{-5}$, (E) SKCM, CCL5, $P = 2.65 \times 10^{-5}$, (F) SKCM, STAT4, $P = 2.84 \times 10^{-7}$, (G) SKCM, XCL2, $P = 2.27 \times 10^{-6}$, (H) SKCM, Srgn, $P = 6.10 \times 10^{-7}$.

discriminate ALS patients from healthy controls with something more easily obtained. For example, blood microRNA can be one such alternative biomarker that can discriminate ALS patients from healthy controls. Generally, microRNAs are supposed to be more disease specific than mRNA and are frequently employed as disease biomarkers. In addition to this, in contrast to tissue mRNA, blood microRNA can be obtained relatively in a less invasive manner.

In order to see if blood microRNA can be an alternative biomarker that can discriminate ALS patients from healthy controls, we consider serum microRNAs composed of 53 expression profiles (details are described in Materials and Methods). At first, we apply PCA to 53 miRNA expression profiles such that PC loading, v_l, are attributed to samples (healthy controls and patients). Then we find that v_2 is associated with BH criterion adjusted P-values less than 0.05, which is computed by categorical regression (ANOVA) assuming four classes: (1) healthy controls, (2) sALS patients, (3) fALS patients, and (4) mutation carriers without symptoms. Then we attribute P-values to miRNAs by assuming that u_2 obeys χ^2 distribution; obtained P-values are adjusted by BH criterion. Finally, as many as 107 miRNAs are associated with adjusted P-values less than 0.2 (full list of selected 107 miRNAs is available as Table S3 [3]).

In order to see if selected miRNAs are reasonable, we try to discriminate four classes using selected 107 miRNAs. PC loading, v_l', is re-computed by applying PCA to only these 107 miRNAs selected. In order to select PC loadings that are distinct between four classes, we apply categorical regression to re-computed PC loading, v_l'.

$$v'_{lj} = a_l + \sum_s b_{ls} \delta_{sj}$$

where δ_{sj} takes 1 only when j belongs to the sth category, otherwise it takes 0, and a_l and b_{ls} are regression coefficients. P-values are attributed to v_l's and corrected by BH criterion. The first, second, third, and eighth PC loading (Fig. 2) are associated with adjusted P-values less than 0.05. The 53 serum microRNA profiles composed of 4 classes are discriminated by LDA using the 4 PC loadings (Table 2, Fig. 3). Although accuracy is as small as 66%, it is still valid since expected accuracy is as small as 0.25 because of having 4 categories (P-values computed by χ^2 test is 7.05×10^{-9}). Thus, the identified 107 serum microRNAs have clear ability to discriminate 4 categories effectively.

Although it is obvious that the selected 107 microRNAs are effective biomarkers for ALS patients, we are not sure if the

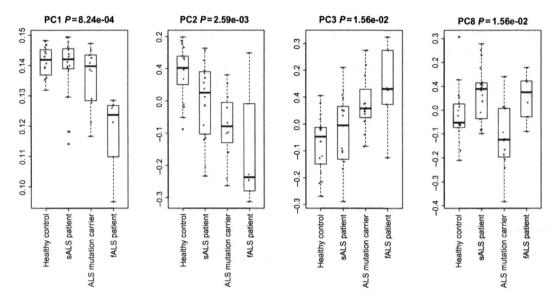

Fig. 2 Boxplots of the first, second, third, and eighth PC loadings attributed to samples composed of four categories: (1) healthy controls, (2) sALS patients, (3) ALS mutant carriers, and (4) fALS patients. P-values are based upon categorical regression (not corrected).

Table 2 Confusion matrix of four ALS categories.

Prediction	True			
	Healthy controls	sALS patients	ALS mutation carriers	fALS patients
Healthy controls	14	6	0	0
sALS patients	3	8	2	0
ALS mutation carriers	0	2	8	1
fALS patients	0	2	2	5

selected 107 microRNAs are also biologically reasonable. One possible validation is to see if microRNAs selected are up/downregulated between ALS patients and healthy controls. In order to determine this, we compare 107 microRNAs with 33 microRNAs previously identified as downregulated ones [11]. Then we find as many as 27 significant interactions between 107 microRNAs and 33 microRNAs. We also compare

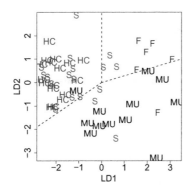

Fig. 3 Scatter plot of 53 miRNA profiles based upon the first and second LDs. *HC*, healthy control; *MU*, mutation carrier; *F*, fALS; *S*, sALS.

107 miRNAs with 38 downregulated microRNAs in ALS [12] and find 16 overlapped microRNAs. These suggest that our strategy successfully identifies microRNAs whose expression is altered by ALS disease.

This is only validation for downregulated miRNA. It is better to have validations for upregulated miRNAs. Since we have four classes, it is not easy to select upregulated miRNAs. Thus we have done the following. In LDA, LD function can be composed of linear combinations of 107 miRNAs. We notice that 27 downregulated miRNAs have larger contributions to the first and the second LD. Then we selected 24 miRNAs associated with smaller contributions to the first and the second LD as upregulated miRNAs.

In order to validate these 24 miRNAs biologically, we uploaded these 24 upregulated microRNAs as well as 27 downregulated microRNAs to DIANA-mirpath [13], which performs enrichment analysis based upon mRNAs targeted by uploaded miRNAs. The enrichment analysis for 27 downregulated microRNAs is shown in Table 3. The enrichment analysis for 24 upregulated microRNAs is available as Table S2 [3].

Some of the 19 KEGG pathways in Table 3 were previously reported to be related to ALS. For example, two KEGG pathways, "extracellular matrix (ECM)-receptor interaction" and "focal adhesion," were reported to be expressed differentially between ALS patients and healthy controls [14]. On the other hand, "adherens junction" protein E-cadherin is downregulated in an ALS mouse model [15]. These are only a few examples. Regardless, from the biological point of view, the selected 107 miRNAs are reasonable.

Table 3 KEGG pathway enrichment analysis for 27 downregulated microRNAs by DIANA-mirpath.

KEGG pathway	P-value	#Genes	#miRNAs
ECM-receptor interaction	1.45×10^{-10}	10	3
Adherens junction[a]	1.21×10^{-9}	16	6
Transcriptional misregulation in cancer[a]	4.39×10^{-6}	23	6
Cell cycle[a]	4.39×10^{-6}	25	8
Hippo signaling pathway[a]	6.77×10^{-6}	24	9
Oocyte meiosis[a]	3.39×10^{-5}	18	5
TGF-beta signaling pathway[a]	0.000314	13	2
Protein processing in endoplasmic reticulum[a]	0.000971	25	8
Pantothenate and CoA biosynthesis	0.001459	3	5
RNA transport[a]	0.002423	23	7
Focal adhesion[a]	0.005153	29	8
Ubiquitin mediated proteolysis[a]	0.008498	21	8
Colorectal cancer[a]	0.010697	10	7
2-Oxocarboxylic acid metabolism	0.017091	3	3
mRNA surveillance pathway[a]	0.023398	15	4
AMPK signaling pathway[a]	0.023398	19	6
Proteoglycans in cancer[a]	0.023929	23	6
Regulation of actin cytoskeleton[a]	0.03352	23	6
Spliceosome	0.03352	18	7

[a]Denotes that these pathways are targeted by both up- and downregulated miRNAs.

3.3 Substantia nigra mRNA expression profiles for PD patients

In the previous two sections, we successfully identify mRNAs as well as microRNAs as biomarkers for ALS. Nevertheless, if we can identify microRNA biomarkers only from mRNA measurement, it is very useful. Biologically, it should be possible. As shown in the previous sections (Table 3), identified microRNA biomarkers have biologically reasonable target mRNAs. Conversely, microRNAs that target mRNAs identified as biomarkers should be predicted staring from the list of mRNAs.

In order to see if it is possible, we consider here AD to identify biomarkers and try to infer biomarker microRNAs as those targeting identified mRNA. Substantia nigra mRNA expression profiles are downloaded from three independent GEO data sets and merged into one data set with 57 samples composed of 27 healthy controls and 30 PD patients. PCA is applied to the merged data set such that PC loading, v_l, are attributed to samples. After applying t-test to obtained PC loading, v_l, that is composed of two categories (healthy controls and PD patients), P-values are attributed to 57 samples; obtained P-values are corrected by BH criterion, and the sixth PC loading, v_6, is associated with adjusted P-value less than 0.01. Then P-values are attributed to mRNAs assuming χ^2 distribution for PC score, u_6; P-values are corrected by BH criterion and as a result, 255 probes are identified as those associated with corrected P-values less than 0.01. Using these 255 probes, PC loadings, v_l', are recomputed. P-values attributed to v_l' are again computed with t-test and the fourth re-computed PC loading, v_4', is associated with adjusted P-values less than 0.01.

LDA is applied to 57 samples using the fourth re-computed PC loading, v_4' (Table 4). The accuracy is as great as 0.88, which is significant enough (P-value computed by Fisher's exact test is 2.65×10^{-6} and odds ratio is 20.5). Thus, empirically, mRNAs that can discriminate PD patients from healthy controls are successfully identified.

In order to validate selected mRNAs biologically, we identify 244 gene symbols associated with these 255 probes and upload 244 gene symbols to Enrichr. There are many categories enriched with uploaded 244 gene symbols in Enrichr (full list is available as Table S4 [4]). First, the four top ranked biological terms in "Disease Perturbations from GEO down" of Enrichr are PD (Table 5). This suggests that 244 gene symbols associated with the selected 255 probes are primarily related to PD.

Although we are not willing to discuss all of PD-related enrichments found in the original study [4], there was an enormous number of enrichments related to PD.

Next, in order to see if microRNA biomarkers can be identified with these 244 gene symbols, we check "miRTarBase 2017" in Enrichr. We found 15 microRNAs that significantly target 113 gene symbols out of 144 gene symbols uploaded (Table 6)

In order to determine if these 15 microRNAs are biologically reasonable biomarkers, we performed a literature search to see if

Table 4 Confusion matrix of PD patients and healthy controls.

Predict	True	
	Control	PD
Control	24	8
PD	3	22

Table 5 Top five ranked terms of "disease perturbations from GEO down" in Enrichr.

Term	Overlap	P-value	Adjusted P-value
Parkinson's disease DOID-14330 human GSE19587 sample 740	65/207	5.02×10^{-83}	4.18×10^{-80}
Parkinson's disease DOID-14330 human GSE19587 sample 1080	56/167	5.88×10^{-73}	1.60×10^{-70}
Parkinson's disease DOID-14330 human GSE19587 sample 496	73/361	3.90×10^{-78}	1.59×10^{-75}
Parkinson's disease DOID-14330 human GSE7621 sample 940	67/365	2.96×10^{-68}	6.06×10^{-66}
Dystonia C0393593 human GSE3064 sample 329	62/317	1.06×10^{-64}	1.74×10^{-62}

Table 6 15 miRNAs that target 113 gene symbols out of 144 gene symbols uploaded in "miRTarBase 2017" of Enrichr.

Term	Overlap	P-value	Adjusted P-value	Reference
hsa-miR-92a-3p	37/1404	1.41×10^{-8}	2.71×10^{-5}	[16]
hsa-miR-16-5p	37/1555	1.93×10^{-7}	1.85×10^{-4}	[17]
hsa-miR-615-3p	25/891	1.38×10^{-6}	8.85×10^{-4}	[18]
hsa-miR-877-3p	19/606	5.92×10^{-6}	2.28×10^{-3}	[19]
hsa-miR-100-5p	12/250	5.37×10^{-6}	2.28×10^{-3}	[20]
hsa-miR-320a	18/584	1.33×10^{-5}	4.25×10^{-3}	[21]
hsa-miR-877-5p	11/235	1.68×10^{-5}	4.63×10^{-3}	[19]
hsa-miR-23a-3p	11/249	2.88×10^{-5}	6.91×10^{-3}	[20]
hsa-miR-484	22/890	4.37×10^{-5}	9.33×10^{-3}	[20]
hsa-miR-23b-3p	12/322	6.55×10^{-5}	1.26×10^{-2}	[22]
mmu-miR-15a-5p	15/499	9.42×10^{-5}	1.65×10^{-2}	[20]
hsa-miR-324-3p	12/338	1.04×10^{-4}	1.66×10^{-2}	[23]
mmu-miR-19b-3p	11/310	2.03×10^{-4}	3.00×10^{-2}	[24]
mmu-miR-7b-5p	13/438	3.13×10^{-4}	4.02×10^{-2}	[24]
hsa-miR-505-3p	9/222	2.93×10^{-4}	4.02×10^{-2}	[25]

previous studies report the relationship between these microRNAs and PD. All 15 microRNAs are associated with the previous study that reports the relationship with PD. In conclusion, we can successfully identify possible microRNA biomarkers that are reported to be related to PD, starting with mRNAs that are biologically reasonable, as well as discriminate PD patients from healthy controls.

4 Discussions

In this chapter, we selected possible biomarker mRNAs or microRNAs using PCAUFE. As target diseases, we employed two neurodegenerative diseases, ALS and PD. Selected mRNAs and microRNAs are not only biologically reasonable, but also can discriminate healthy controls from patients. In addition to this, candidate microRNAs biomarkers that are more convenient than mRNA because of their total small numbers were shown to be inferred from mRNAs identified. Generally, identification of biomarkers or neurodegenerative diseases are not easy tasks. Successfully application of the recently proposed PCAUFE is promising and is expected to be useful to identify biomarkers for other neurodegenerative diseases.

References

[1] Y.-H. Taguchi, Unsupervised Feature Extraction Applied to Bioinformatics, Springer, 2019.
[2] Y.-H. Taguchi, H. Wang, Genetic association between amyotrophic lateral sclerosis and Cancer, Genes 8 (2017) 243.
[3] Y.-H. Taguchi, H. Wang, Exploring microRNA biomarker for amyotrophic lateral sclerosis, Int. J. Mol. Sci. 19 (2018) 1318.
[4] Y.-H. Taguchi, H. Wang, Exploring microRNA biomarkers for Parkinson's disease from mRNA expression profiles, Cell 7 (2018) 245.
[5] Y. Benjamini, Y. Hochberg, Controlling the false discovery rate: a practical and powerful approach to multiple testing, J. R. Stat. Soc. Ser. B 57 (1) (1995) 289–300.
[6] N. Kolesnikov, E. Hastings, M. Keays, O. Melnichuk, Y.A. Tang, E. Williams, M. Dylag, N. Kurbatova, M. Brandizi, T. Burdett, K. Megy, E. Pilicheva, G. Rustici, A. Tikhonov, H. Parkinson, R. Petryszak, U. Sarkans, A. Brazma, ArrayExpress update—simplifying data submissions, Nucleic Acids Res. 43 (2015) D1113–D1116.
[7] E. Clough, T. Barrett, The gene expression omnibus database, Methods Mol. Biol. 1418 (2016) 93–110.
[8] D.M. Freedman, R.E. Curtis, S.E. Daugherty, J.J. Goedert, R.W. Kuncl, M.A. Tucker, The association between cancer and amyotrophic lateral sclerosis, Cancer Causes Control 24 (2013) 55.
[9] J. Anaya, OncoLnc: linking TCGA survival data to mRNAs, miRNAs, and lncRNAs, PeerJ Comput. Sci. 2 (2016) e67.
[10] K. Tomczak, P. Czerwińska, M. Wiznerowicz, The cancer genome atlas (TCGA): an immeasurable source of knowledge, Contemp. Oncol. (Pozn) 19 (1A) (2015) A68–A77.
[11] A. Freischmidt, K. Muller, L. Zondler, P. Weydt, A.E. Volk, A.L. Bozic, M. Walter, M. Bonin, B. Mayer, C.A. von Arnim, M. Otto, C. Dieterich, K. Holzmann, P.M. Andersen, A.C. Ludolph, K.M. Danzer, J.H. Weishaupt, Serum microRNAs in patients with genetic amyotrophic lateral sclerosis and pre-manifest mutation carriers, Brain 137 (11) (2014) 2938–2950.
[12] M. Liguori, N. Nuzziello, A. Introna, A. Consiglio, F. Licciulli, A.E.D. E. Scarafino, E. Distaso, I.L. Simone, Dysregulation of microRNAs and target

genes networks in peripheral blood of patients with sporadic amyotrophic lateral sclerosis, Front. Mol. Neurosci. 11 (2018) 288 (online).
[13] I.S. Vlachos, K. Zagganas, M.D. Paraskevopoulou, G. Georgakilas, D. Karagkouni, T. Vergoulis, T. Dalamagas, A.G. Hatzigeorgiou, DIANA-miRPath v3.0: deciphering microRNA function with experimental support, Nucleic Acids Res. 43 (W1) (2015) W460–W466.
[14] M.K. Kotni, M. Zhao, D.Q. Wei, Gene expression profiles and protein-protein interaction networks in amyotrophic lateral sclerosis patients with C9orf72 mutation, Orphanet. J. Rare Dis. 11 (2016) 148.
[15] S. Wu, J. Yi, Y.G. Zhang, J. Zhou, J. Sun, Leaky intestine and impaired microbiome in an amyotrophic lateral sclerosis mouse model, Physiol. Rep. 3 (2015) e12356.
[16] P. Chatterjee, M. Bhattacharyya, S. Bandyopadhyay, D. Roy, Studying the system-level involvement of microRNAs in Parkinson's disease, PLoS One 9 (2014) e93751.
[17] A.G. Hoss, A. Labadorf, T.G. Beach, J.C. Latourelle, R.H. Myers, MicroRNA profiles in Parkinson's disease prefrontal cortex, Front. Aging Neurosci. 8 (2016) 36.
[18] A.G. Hoss, V.K. Kartha, X. Dong, J.C. Latourelle, A. Dumitriu, T.C. Hadzi, M.E. Macdonald, J.F. Gusella, S. Akbarian, J.F. Chen, Z. Weng, R.H. Myers, MicroRNAs located in the Hox gene clusters are implicated in huntington's disease pathogenesis, PLoS Genet. 10 (2014) e1004188.
[19] C.R. Sibley, Y. Seow, H. Curtis, M.S. Weinberg, M.J. Wood, Silencing of Parkinson's disease-associated genes with artificial mirtron mimics of miR-1224, Nucleic Acids Res. 40 (2012) 9863–9875.
[20] L. Chen, J. Yang, J. Lu, S. Cao, Q. Zhao, Z. Yu, Identification of aberrant circulating miRNAs in Parkinson's disease plasma samples, Brain Behav. 8 (2018) e00941.
[21] S.M. Heman-Ackah, M. Hallegger, M.S. Rao, M.J. Wood, RISC in PD: the impact of microRNAs in Parkinson's disease cellular and molecular pathogenesis, Front. Mol. Neurosci. 6 (2013) 40.
[22] P. Prajapati, L. Sripada, K. Singh, K. Bhatelia, R. Singh, R. Singh, TNF-α regulates miRNA targeting mitochondrial complex-I and induces cell death in dopaminergic cells, Biochim. Biophys. Acta Mol. basis Dis. 1852 (2015) 451–461.
[23] A. Vallelunga, M. Ragusa, S. Di Mauro, T. Iannitti, M. Pilleri, R. Biundo, L. Weis, C. Di Pietro, A. De Iuliis, A. Nicoletti, Identification of circulating microRNAs for the differential diagnosis of Parkinson's disease and multiple system atrophy, Front. Cell. Neurosci. 8 (2014) 156.
[24] L. Leggio, S. Vivarelli, F. L'Episcopo, C. Tirolo, S. Caniglia, N. Testa, B. Marchetti, N. Iraci, microRNAs in Parkinson's disease: from pathogenesis to novel diagnostic and therapeutic approaches, Int. J. Mol. Sci. 18 (2017) 2698.
[25] S.K. Khoo, D. Petillo, U.J. Kang, J.H. Resau, B. Berryhill, J. Linder, L. Forsgren, L.A. Neuman, A.C. Tan, Plasma-based circulating microRNA biomarkers for Parkinson's disease, J. Parkinsons Dis. 2 (2012) 321–331.

Disease diagnosis using machine learning: A comparative study

Rakshit Jain, Asmita Chotani, and G Anuradha
Vellore Institute of Technology, Vellore, India

1 Introduction

Disease diagnosis is the system for determining the disease based on the symptoms the patient inputs. It is important since some symptoms are unnoticeable and tend to get overlooked like the redness of the eyes, considering the patient might have rubbed his/her eyes, but redness of the eyes can be a caused due to serious issues like viral or glaucoma as well. People tend to avoid going to the doctors in such scenarios thinking of it as a temporary problem. It is quite natural because regular visits to the doctors for even the slightest symptoms not only would be costly but also will waste time of both the patient and the doctor. At other times the patient is not able to detect the seriousness of the symptoms in time, weight loss is one such symptom. Weight loss is a symptom that can be misunderstood as the result of positive impact on the body, but the reason behind it can also be negative; it can be a result of diseases like diabetes, jaundice, tuberculosis, and hyperthyroidism. Therefore, if the real cause behind weight loss is not determined on time, there is a high risk of the disease severity increasing and putting the sufferer's life in danger. The proposed system solves the aforementioned problem of whether to go to the doctor or not by bringing the doctor's diagnosis to the patient. The patient would be able to check for the slightest changes in their health without actually consulting with the doctor. For the proposed system a dataset consisting of various symptoms and the diseases diagnosed from each is taken into consideration, which is created with the consultation of doctors specializing in various domains. Two machine learning model is trained using the dataset to analyze and diagnose the disease. With the technology enhancing with time, it has become easier for performing analysis and predicting things that took a lot of research before. The first model, the decision tree algorithm, is the most common and useful machine learning algorithm for classification. The decision tree thus works on each and every combination to classify the symptoms entered into clusters of symptoms that cause a similar disease. A number of variations of decision trees are created by varying the tree depth and the number of leaf nods for the various symptoms. These variations are then compared to find out the most accurate decision tree among all. Data visualization techniques are used to compare the accuracy for various tree depths and number of leaf nodes. The second model of deep neural networks makes use of a number of hidden layers to get an accurate result. While the DNN model has been used by many to predict or diagnose chronic diseases, not much work has been done using it to predict acute diseases, which our proposed DNN model will fulfill. The proposed models are created for about 41 diseases and 132 symptoms for the diagnoses of the diseases. The diseases include acute diseases like common cold and pneumonia and chronic diseases like various types of hepatitis, AIDS, impetigo, and jaundice. High fever is a common symptom for 12 of the considered 40 diseases. This makes it difficult for the user to actually guess the disease themselves; therefore the proposed system will take each symptom entered by the user into consideration and display the most probable disease. For example, if the user enters the symptoms as chills, fatigue, high fever, sunken eyes, sweating, and fast heart rate, the proposed model will provide the most diagnosed disease, which in this case is pneumonia, along with the probability of the other 39 disease as well.

This chapter addresses the following points:
- Acute and chronic diseases like dengue, diabetes, peptic ulcer, gastroenteritis, osteoarthritis, and hypoglycemia can be easily diagnosed from the proposed model.
- The symptoms that the user chooses from are common terms and easily understandable by the normal man by our proposed system. Some of the symptoms include nausea, vomiting, acidity, and joint pain.
- This developed model will help keep one's health in track. The proposed system is basic and can be easily understood by nontechnical and nonmedical people.

2 Related work

Several ways have been considered when it comes to collecting data and predicting the disease. Wearable sensors were one such way to collect data to train the model to detect Parkinson's disease [1]. Research on the topic has concluded that by using the wearable sensors, the machine learning algorithm can automatically detect Parkinson's disease and even detect its progression. Convolutional neural network has been used to detect the effect of various movements on Parkinson's disease. The research paper compared repeated assessments of the same movements to detect the disease. Researchers have been interested in applying machine learning to healthcare systems lately. Research has been done to provide a solution such that the symptoms are detected from the faces to diagnose various visually observable diseases [2]. The solution put forward involves analyzing various common features of the face, detecting and categorizing the illness features, and then testing the user's facial features with the trained model. SVM classifier and PCA has been used to extract the phenotypic information from ordinary photographs of the patients, be it clinical or nonclinical, and model the dysmorphisms of the human face in a multidimensional space, which the research stated as a "Clinical Face Phenotype Space" [3]. It uses image processing techniques to detect the face and further detect feature point annotation. Considering the death rate of children in Ghana, researchers presented ways to identify the symptoms of diseases, which the children mainly suffer from in Ghana [4].

The research talked about the mobile phone being used as a tool to diagnose symptoms like cough, fever, vomiting, cold, and diarrhea. Research has been done by researchers to compare various algorithms that have been used till date to detect the probability for the disease that are diagnosed [5]. The algorithms compared and analyzed are Naïve Bayes and Apriori Algorithms. These supervised learning algorithms were used to diagnose the disease. The results for both the algorithms were analyzed to infer which of the two algorithms was better. Artificial intelligence and neural networks have also been used to create a disease detection system [6]. Researchers have worked upon such systems using data mining techniques and further used techniques like SVM, Naïve Bayes, decision trees, clustering, and regression to analyze the medical data and to detect swine flu based on the symptoms, laboratory indicators, and many other indicators. During initial days of machine learning trend, research was done on various techniques of machine learning including Naïve Bayes, KNN, K-means, principal component analysis, and linear discriminant analysis to reduce the dimensions [7]. Research papers have been written to explain about the various classifying ad reduction algorithms. Researchers have worked upon applying latent factor models to use machine learning to diagnose the disease even if data are somewhat incomplete [8]. The algorithm proposed by the researcher used structured and unstructured data to apply convolutional neural network and predict the probability of the disease. Comparative analysis of machine learning algorithms has been done when it comes to diagnosing diseases like heart disease, dengue, diabetes, and liver diseases [9]. The various machine learning techniques taken into consideration were supervised learning, deep learning, unsupervised learning, reinforcement learning, evolutionary learning, and semi supervised learning. The best algorithm was determined for each type of disease—Naïve Bayes being the best for diabetes, SVM for heart disease was inferred as the result of the research paper. The most common way chosen by researchers has been to obtain the data from past patient cases. The researchers made use of information about these cases as knowledge and represented the knowledge with the help of production rules and neural networks [10]. Researchers created an expert system to analyze laboratory exams and list all the possible diseases that can be diagnoses to help physicians. Work has also been done in creating expert systems to diagnose the disease and providing with methods of treatment to the user. Researchers created an expert system DExS [11], that is, disease expert system that makes use of inference rules to diagnose the disease and provide methods of treatment. One of the new techniques being researched upon is deep neural networks. Researchers have studied the DNN method to detect objects and classify them based on the location and class (e.g., table and cow). The detection has been done using DNN regression. The model studied by the researcher had seven layers, which were responsible for convolution, generating object binary mask, which detected the object. To detect complex objects and those in contact with each other, the researchers used five masks to differentiate the boundaries [12]. Further deep convolutional neural network was studied for the process of image deconvolution, which is responsible to restore images. The process studied combined two muddles, one for deconvolution CNN and one for denoise CNN. With the help of a number of layers, the model was able to sharpen a blur image and obtain a clear image [13]. There are plenty of other applications that can be found on machine learning and deep learning or LASSO linear regression model [14, 15] be it classifying spam emails [16] and stock market forecasting [17]. Deep neural networks have also been used to create a model for stock market forecasting. With a configuration of (20-18-21-1), the model used the ReLU function and the sigmoid function to activate the input and output nodes, and the model wasted less computational time by avoiding pre training of the data. It used the adaptive learning rate algorithm and overfitting techniques to obtain suitable results [18]. Some other applications of machine learning and deep learning that have been researched about include intrusion detection systems that classify intrusion attacks using stacking classification [19], deep neural network and support vector machine [20], predicting models to predict load both heating and cooling of residential buildings [21] using various techniques like multivariate adaptive regression splines (MARS), extreme learning machine (ELM) and a hybrid model of both [22], study of factors of nuclear reprogramming for generation of induced pluripotent stem (iPS) cells [23]. While regression techniques were applied by researchers to predict the capacity of suction caisson to uplift [24], deep learning was used to predict seismic events [25].

Numerous researches have been done in the field of medicine using different techniques; researchers have analyzed cancer data to estimate the tolerance factor of people based on age using variable precision rough set [26]; image processing and machine learning techniques have been used to segment retinal blood vessels to further diagnose diabetic retinopathy using fundus images [27] and ant colony optimization technique to predict heart diseases [28]. Convolutional neural network, a popular technique to find the solution to complex problems [29], has been used by researchers to make use of MRI images and various other medical images to recognize and detect Alzheimer's disease [30], osteoarthritis [31], and prognosis of skin lesions [32] (Table 1).

3 Methodology

In this section the implemented methods along with some basic knowledge of techniques used and information related to them will be discussed.

3.1 Decision tree model

3.1.1 Decision tree algorithm

The Decision tree algorithm is the most used algorithm to represent decision making and classification. The algorithm analyzes the data and finds out different features and conditions that are used to split the data into multiple classes. Being a supervised

Table 1 Summary of recent work based on disease diagnosis.

Authors	Model used	Dataset used	Target attribute
Swapna G., Vinayakumar R., Soman K.P. [33]	LSTM, CNN	Electrocardiograms of people form normal and diabetes group	Diabetes prediction
Timothy J. Wroge [34]	Deep neural networks	mPower voice dataset	Parkinson's disease diagnosis
Youness Khourdifi, Mohamed Bahaj [35]	KNN, SVM, Naïve Bayes, random forest, multilayer perception, PSO, ACO	Heart disease dataset (UCI repository)	Heart disease prediction classification
Vinay Shekhar Bannihatti Kumar, Sujay S. Kumar, Varun Saboo [36]	KNN, decision tree, neural network	UCI dataset	Dermatological disease detection
K. Vembandasamy, R. Sasipriya, E. Deepa [37]	Naïve Bayes	Diabetes dataset (500 patients)	Heart disease detection
Li-Sheng Wei, Quan Gan, Tao Ji [38]	Gray level cooccurrence matrix, support vector machine	Skin disease dataset (herpes, dermatitis, psoriasis, etc.)	Skin disease recognition
Igor Kononenko [39]	Naïve Bayes, neural networks	Eight different medical datasets	Comparison between machine learning techniques for diagnosis

learning algorithm, decision tree algorithm requires being trained and tested by the same dataset. To create a strong classifier model, boosting is done on weak classifier models. The boosting process includes two models: The first model is built from the training data, and the second is built to overcome the errors and correct them in the first model. Boosting process keeps on adding new models until the data are predicted most accurately and precisely. The basic decision tree algorithm used has been described by Hu [40] and is given in the succeeding text.

Algorithm 1 Decision tree method [40]

```
DecisionTreeAlgorithm(TrainingData, Goal, Attribute)
    TrainingData: dataset used for training to fit
parameters
    Goal: target feature to be evaluated i.e. Prognosis
    Attribute: set of properties for analysis
    {
```

```
  Create Root node
  If TrainingData have same Goal value g_i,
     Then Return tree with single node, i.e. Root, with
Goal = g_i
  If Attribute = empty (i.e. there is no properties
present ),
     Then Return tree with single node, i.e. Root, with
most modal value of goal in TrainingData
        Otherwise
  {
  Choose feature X which most aptly classifies Trai-
ningData using entropy measure
  Set X as attribute being used for root
  For each legal value of X, y_i, do
  {
  Add branch corresponding to X = y_i
  Let TrainingData_{y_i} be subset of TrainingData that have
X = y_i
  If TrainingData_{y_i} is empty,
       Then add node where Goal value = most modal value
of Goal in TrainingData
          Else
     add subtree DecisionTreeAlgorithm(TrainingData,
Goal, Attribute-{X})
  }
  }
  Return (Root)
  }
```

3.1.2 AdaBoostClassifier

AdaBoostClassifier is the most basic and general boosting method. It is used for enhancing the performance of classification of data. It calculates the average of the various weak classifier models to make predictions. The best classifier model that goes along well with AdaBoostClassifier is the decision tree model. The algorithm for AdaBoostClassifier has been explained by Schapire [41] along with its various applications. The AdaBoost Algorithm is given in the succeeding text.

Algorithm 2 AdaBoost algorithm [41]

```
Provided elements for training : (z_j,w_j)
    where j ranges from 1 to q, {(z_1,w_1),.........,(z_q,
w_q)}
       w_j takes values -1 or +1
       zj has domain β
```

```
Step1- 'X' Distributions are created for all
rounds(Dist₁....Distₓ)
       Such that Dist₁(j)=1/q where j ranges from 1 to q
  Distributions created are-
       Dist₁(1)......Dist₁(q)........Dist₂(q)
Step 2- Use above Distributions to train weak trees
Step 3- Obtain hypothesis 'hₓ' with error 'err' such
that error is less-
   errₓ=probability[hₓ(zⱼ)≠wⱼ]
Step 4- Let αₓ = ln(1-errₓ / errₓ)
Step 5- Update Distribution value
   Dist ₓ₊₁(j) = Distₓ(j)exp(-αₓwⱼhₓ(zⱼ)) / Nₓ
   Where Nₓ is for normalizing Distribution
Step 6- Final Hypothesis (combination of all hypothesis
   H(z)=∑ˣₓ₌₁αₓhₓ(z)
```

3.1.3 Implemented method

Classification of various symptoms is done on the basis of the diseases diagnosed. The machine learning model created for the classification makes use of decision trees and the AdaBoostClassifier. The AdaboostClassifier is in general mainly used for boosting the classification process. The proposed model uses the DecisionTreeClassifier function to create the various decision trees, and further the AdaBoostClassifier takes into consideration a number of weak decision tree to build a strong one. Both the training and testing databases consist of the symptoms for various diseases. The decision tree model is created using the DecisionTreeClassifier function and the training dataset. Executing the prototype at a small scale, the accuracy of the decision tree model is found by the process of testing, which is done by using the testing dataset. Data visualization techniques including plotting of graphs, and plots are used to compare the accuracies and clearly depict the comparisons to the user. Graphs are plotted for symptoms and diseases to clearly portray how common the particular symptom and disease are. The graphs can warn the users beforehand about the commonness of the symptoms. The diagnosis process involves the user entering the symptoms that are then classified based on the decision tree model created. For the input procedure the various related symptoms of the dataset are displayed in groups of 10 out of which the user is required to enter the symptoms he/she is suffering from. All the entered symptoms are then appended into an array. To pass the user's input in the decision tree model, the array created is converted into a sparse array (consisting of 1's and 0's). The model then classifies the symptoms into clusters of various diseases (with related symptoms). As a result the model displays the most probable disease that can be diagnosed from the entered symptoms. A number of libraries have been used to carry out the development of the proposed model.

3.2 Deep neural network method

3.2.1 Activation functions

Activation functions are responsible for calculating the sum of the product of the various weights and inputs with the bias to determine the final output value for the current hidden layer, which would be the input for the next layer [42]. The two activation functions used in the proposed model are as follows (Fig. 1):

(a) *Softmax*

The Softmax activation function normalizes the input values such that all the output values lie between 0 and 1, with their sum equal to 1. It determines the probability of each class being true or false.

If Input is $x = \begin{bmatrix} 1.1 \\ 0.8 \\ 0.6 \end{bmatrix}$, then $F(x) = \begin{bmatrix} 0.44 \\ 0.32 \\ 0.24 \end{bmatrix}$, that is, if x is the vector of inputs, the softmax function $F(x)$ is $F(x_j) = \dfrac{x_j}{\sum_{i=1} x_i}$ where x_j is the jth value of the vector.

(b) *Rectified linear unit (ReLU)*

It is a piecewise linear function with the following function used to obtain the summed value of the inputs:

$$F(x) = \begin{cases} x & \text{if } x \text{ is positive} \\ 0 & \text{otherwise} \end{cases}$$

3.2.2 Implemented DNN method

The DNN model implemented made use of for a number of hidden layers to predict the disease with the input given as various symptoms. The model had a fixed number of 133 nodes in input layer since the dataset contains 133 features and 41 nodes in output layer because of 41 unique diseases in dataset. The input would be of the form 0/1; the input will be "1" only if the symptom is present. Each hidden layer was developed to have 87 nodes, and the prediction was checked for 0–5 number of hidden layers. The batch size for the model has been taken as 2000; therefore for each

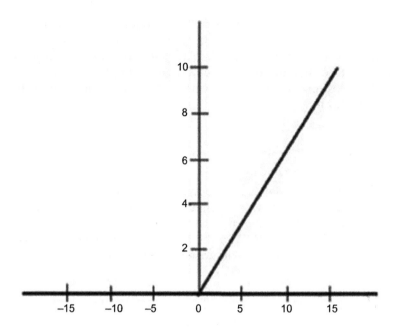

Fig. 1 ReLU activation function.

Table 2 Architecture of DNN.

Layers	Model Units/nodes	Activation
Input	133	ReLU
Hidden layer 1	87	ReLU
Hidden layer 2	87	ReLU
Output	41	Softmax

epoch after reading 2000 data entries, the model updates the weights before moving on to the next entries, and this procedure is continued until the whole dataset has been processed. The method uses the fully connected sequential model to get the desirable structure of the network [43–45]. The architecture of the model has been shown in Table 2, specifying the activation functions and the number of nodes for each layer of the structure. Further the structure of the DNN model has been portrayed in Fig. 2.

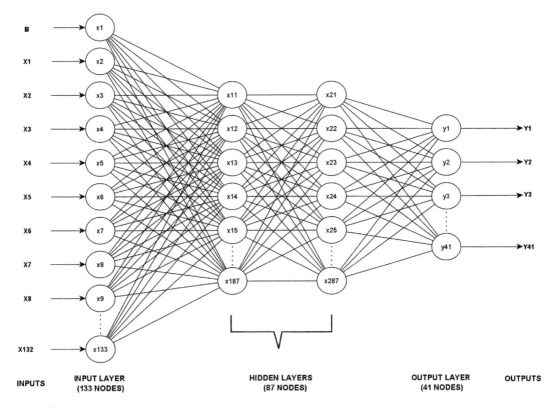

Fig. 2 DNN model.

4 Dataset and metrics

Two different datasets were used to train and test the models. The datasets were created with modification in disease diagnosis datasets available online to make them suitable for the task. The training datasets consisted of about 5–10 combination of symptoms for various diseases, while the testing dataset consisted of a combination of various diseases and what the prognosis should be, which is compared with the output of the model in the end. Fig. 3 portrays the occurrence of some of the symptoms being considered while training the models. The dataset comprises the information of 132 symptoms, which are treated as the features. The presentation of symptom is as follows:

$$y(\text{Symptom}) = \begin{cases} 1 & \text{Symptom present} \\ 0 & \text{Symptom absent} \end{cases}$$

Prognosis is the output column, which consists of the names of the corresponding disease to each entry of occurrence of various

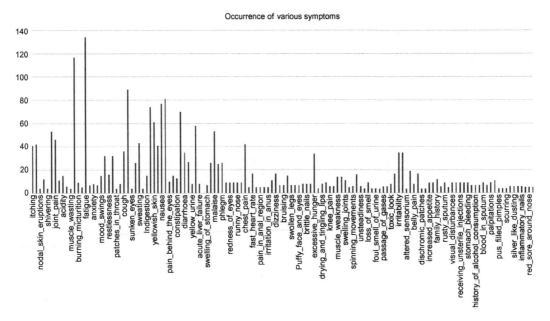

Fig. 3 Occurrences of symptoms considered in training the model.

symptoms. Prognosis contains 41 diseases in total, which results in the model having 41 different classes. Some of the diseases taken into consideration are malaria, dengue, common cold, AIDS, and hypertension.

5 Result

The system will be able to diagnose the disease from the symptoms entered by the users. A model has been trained for 132 symptoms and 41 diseases. The model will be able to successfully diagnose the diseases provided the symptoms and display the most probable disease along with its probability and the probability of various other diseases.

5.1 Comparing accuracy for decision tree model with and without AdaBoostClassifier

From the analysis performed for the variations in decision tree, the best way to represent the model for classification of symptoms based on diseases can be inferred. Fig. 4 displays the graphs plotted to portray information about symptoms and diseases, which

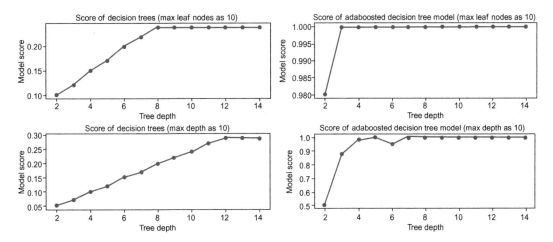

Fig. 4 Variance in accuracy score of decision tree and boosted decision tree.

include the occurrences of both. This reduces the risk of diseases initially being misinterpreted by the user from the visible symptoms. Fig. 4 portrays the accuracy score of the decision tree with and without boosting. The figure can be used to infer that the AdaBoostClassifier increases the accuracy score to the highest it can by combining the weak decision tree models to create a strong one, thus improving the performance.

It can be observed from Fig. 4 that in the model using simple decision tree method with the maximum leaf nodes as 10 and maximum depth as 10, the highest accuracy obtained was of 0.3 while that of the model with AdaBoostClassifier was 1. The highest accuracy was reached at an earlier stage with AdaBoostClassifier, making it the more desirable model among the two.

5.2 Comparing the accuracy for DNN model

Fig. 5 portrays the relation between accuracy and the number of epochs for a specific number of hidden layers. It was observed from the later figure that after considering the number of hidden layers as 1, the DNN model reached its saturation point as there was only a minimal change in its accuracy even after increasing the number of hidden layers. With increase in number of layers, if the epochs are increased, then DNN can provide better results. This means that to obtain the best result for the given dataset, the deep neural networking model should have three layers in total, the input layer, a hidden layer, and an output layer.

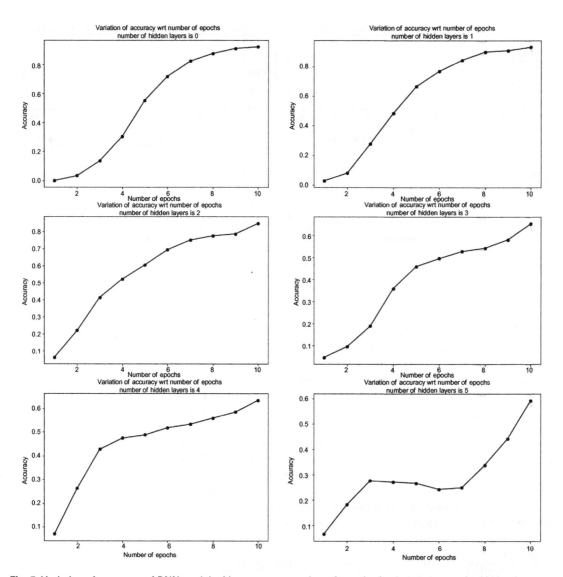

Fig. 5 Variation of accuracy of DNN model with respect to number of epochs for 0, 1, 2, 3, 4, and 5 hidden layers, respectively.

6 Conclusion

The proposed system as a result will thus be able to help people diagnose the disease given the simplest to most complex symptom. This will reduce the time consumption of the user and help in keeping him/her aware about his/her health. Giving people a way to check whether they are suffering from a disease or not, the system would be able to decrease the anxiousness when it comes to being worried about one's health. The accuracy increased with increasing number of leaf nodes and constant tree depth; this increases in accuracy further results in reduction of error rate when using the AdaBoostClassifier. Therefore, with the increase in number of leaf nodes at the max depth, the decision tree model becomes more and more precise. Further, to obtain the best results using DNN Model, a single hidden layer of 83 nodes is sufficient apart from input and output layer.

References

[1] L. Lonini, A. Dai, N. Shawen, T. Simuni, C. Poon, L. Shimanovich, ... A. Jayaraman, Wearable sensors for Parkinson's disease: which data are worth collecting for training symptom detection models. NPJ Digit. Med. 1 (1) (2018), https://doi.org/10.1038/s41746-018-0071-z.

[2] K. Wang, J. Luo, Detecting visually observable disease symptoms from faces. EURASIP J. Bioinform. Syst. Biol. 2016 (1) (2016), https://doi.org/10.1186/s13637-016-0048-7.

[3] Q. Ferry, J. Steinberg, C. Webber, D.R. Fitzpatrick, C.P. Ponting, A. Zisserman, C. Nellåker, Diagnostically relevant facial gestalt information from ordinary photos. eLife 3 (2014), https://doi.org/10.7554/elife.02020.

[4] K.H. Franke, R. Krumkamp, A. Mohammed, N. Sarpong, E. Owusu-Dabo, J. Brinkel, ... B. Kreuels, A mobile phone based tool to identify symptoms of common childhood diseases in Ghana: development and evaluation of the integrated clinical algorithm in a cross-sectional study. BMC Med. Inform. Decis. Mak. 18 (1) (2018), https://doi.org/10.1186/s12911-018-0600-3.

[5] Sunny, A. D., Kulshreshtha, S., Singh, S., Srinabh, Ba, M., H. Sarojadevi, Disease diagnosis system by exploring machine learning algorithms. Int. J. Innov. Eng. Technol., 10(2), 2018.

[6] D. Raval, D. Bhatt, M.K. Kumhar, V. Parikh, D. Vyas, Medical diagnosis system using machine learning. IJSCS 7 (2015) 177–182, https://doi.org/10.090592/IJCSC.2016.026.

[7] S. Razia, P.S. Prathyusha, N.V. Krishna, N.S. Sumana, A review on disease diagnosis using machine learning techniques, Int. J. Pure Appl. Math. 117 (16) (2017) 79–86.

[8] D.K. Harini, M. Natesh, Prediction of probability of disease based on symptoms using machine learning algorithm, Int. Res. J. Eng. Technol. 05 (05) (2018) 392–395.

[9] M. Fatima, M. Pasha, Survey of machine learning algorithms for disease diagnostic. J. Intell. Learn. Syst. Appl. 09 (01) (2017) 1–16, https://doi.org/10.4236/jilsa.2017.91001.

[10] D. Biswas, S. Bairagi, N. Panse, N. Shinde, Disease diagnosis system, Int. J. Comput. Sci. Inform. 1 (2) (2011) 48–51.

[11] P.K. Patra, D.P. Sahu, I. Mandal, An expert system for diagnosis of human diseases, Int. J. Comput. Appl. 1 (13) (2010) 71–73.

[12] C. Szegedy, A. Toshev, D. Erhan, Deep Neural Networks for Object Detection, NIPS, 2013.

[13] L. Xu, J. Ren, C. Liu, J. Jia, Deep convolutional neural network for image deconvolution, Adv. Neural Inform. Process. Syst. 2 (2014) 1790–1798.

[14] S.S. Roy, P. Samui, R. Deo, S. Ntalampiras (Eds.), Big Data in Engineering Applications, In: vol. 44, Springer, 2018.
[15] S.S. Roy, D. Mittal, A. Basu, A. Abraham, Stock market forecasting using LASSO linear regression model, in: Afro-European Conference for Industrial Advancement, Springer, Cham, 2015, pp. 371–381.
[16] S.S. Roy, V.M. Viswanatham, Classifying spam emails using artificial intelligent techniques, Int. J. Eng. Res. Africa 22 (2016) 152–161 Trans Tech Publications.
[17] S.S. Roy, R. Chopra, K.C. Lee, C. Spampinato, B. Mohammadi-ivatlood, Random forest, gradient boosted machines and deep neural network for stock price forecasting: a comparative analysis on south Korean companies, Int. J. Ad Hoc Ubiq. Comput. 33 (1) (2020) 62–71.
[18] D. Shah, W. Campbell, F. Zulkernine, A Comparative Study of LSTM and DNN for Stock Market Forecasting. (2018), https://doi.org/10.1109/BigData.2018.8622462. Schapire, R.E. (2013). Explaining AdaBoost. Empirical Inference.
[19] S.S. Roy, P.V. Krishna, S. Yenduri, Analyzing intrusion detection system: an ensemble based stacking approach, in: 2014 IEEE International Symposium on Signal Processing and Information Technology (ISSPIT). IEEE, 2014, pp. 000307–000309.
[20] S.S. Roy, A. Mallik, R. Gulati, M.S. Obaidat, P.V. Krishna, A deep learning based artificial neural network approach for intrusion detection, in: International Conference on Mathematics and Computing, Springer, Singapore, 2017, pp. 44–53.
[21] S.S. Roy, P. Samui, I. Nagtode, H. Jain, V. Shivaramakrishnan, B. Mohammadivatloo, Forecasting heating and cooling loads of buildings: a comparative performance analysis, J. Ambient Intell. Human. Comput. 10 (2019) 1–12.
[22] S.S. Roy, R. Roy, V.E. Balas, Estimating heating load in buildings using multivariate adaptive regression splines, extreme learning machine, a hybrid model of MARS and ELM, Renew. Sust. Energ. Rev. 82 (2018) 4256–4268.
[23] S.S. Roy, C.H. Hsu, Z.H. Wen, C.S. Lin, C. Chakraborty, A hypothetical relationship between the nuclear reprogramming factors for induced pluripotent stem (iPS) cells generation–bioinformatic and algorithmic approach, Med. Hypotheses 76 (4) (2011) 507–511.
[24] P. Samui, D. Kim, J. Jagan, S.S. Roy, Determination of uplift capacity of suction caisson using Gaussian process regression, minimax probability machine regression and extreme learning machine, Iran. J. Sci. Technol. Trans. Civil Eng. 43 (1) (2019) 651–657.
[25] Y. Geng, L. Su, Y. Jia, C. Han, Seismic events prediction using deep temporal convolution networks. J. Electr. Comput. Eng. 2019 (2019) 1–14, https://doi.org/10.1155/2019/7343784.
[26] S.S. Roy, A. Gupta, A. Sinha, R. Ramesh, Cancer data investigation using variable precision rough set with flexible classification, in: Proceedings of the Second International Conference on Computational Science, Engineering and Information Technology, ACM, 2012, pp. 472–475.
[27] R. Biswas, A. Vasan, S.S. Roy, Dilated deep neural network for segmentation of retinal blood vessels in fundus images, Iran. J. Sci. Technol. Transact. Electr. Eng. (2019) 1–14.
[28] S. Saranya, P. Deepika, D. Sasikala, Comprehensive review on heart disease prediction using optimization techniques. EPRA Int. J. Res. Dev. (2020) 7–13, https://doi.org/10.36713/epra3887.
[29] S. Indolia, A.K. Goswami, S. Mishra, P. Asopa, Conceptual understanding of convolutional neural network- a deep learning approach. Proc. Comput. Sci. 132 (2018) 679–688, https://doi.org/10.1016/j.procs.2018.05.069.

[30] S.S. Roy, R. Sikaria, A. Susan, A deep learning based CNN approach on MRI for Alzheimer's disease detection, Intell. Decis. Technol. (2020) 1–11 (Preprint).

[31] J. Lim, J. Kim, S. Cheon, A deep neural network-based method for early detection of osteoarthritis using statistical data. Int. J. Environ. Res. Public Health 16 (7) (2019) 1281, https://doi.org/10.3390/ijerph16071281.

[32] A.R. Ratul, M.H. Mozaffari, W.-S. Lee, E. Parimbelli, Skin Lesions Classification Using Deep Learning Based on Dilated Convolution. (2019), https://doi.org/10.1101/860700.

[33] G. Swapna, R. Vinayakumar, K.P. Soman, Diabetes detection using deep learning algorithms, ICT Express 4 (4) (2018) 243–246. ISSN 2405-9595, https://doi.org/10.1016/j.icte.2018.10.005.

[34] T.J. Wroge, Y. Özkanca, C. Demiroglu, D. Si, D.C. Atkins, R.H. Ghomi, Parkinson's disease diagnosis using machine learning and voice, in: 2018 IEEE Signal Processing in Medicine and Biology Symposium (SPMB), 2018, pp. 1–7.

[35] Y. Khourdifi, M. Bahaj, Heart disease prediction and classification using machine learning algorithms optimized by particle Swarm optimization and ant colony optimization. Int. J. Intell. Eng. Syst. 12 (2019), https://doi.org/10.22266/ijies2019.0228.24.

[36] V.B. Kumar, S.S. Kumar, V. Saboo, Dermatological disease detection using image processing and machine learning, in: 2016 Third International Conference on Artificial Intelligence and Pattern Recognition (AIPR), Lodz, 2016, pp. 1–6.

[37] K. Vembandasamy, R. Sasipriya, E. Deepa, Heart diseases detection using naive Bayes algorithm, Int. J. Innov. Sci. Eng. Technol. 2 (2015) 441–444.

[38] L. Wei, Q. Gan, T. Ji, Skin Disease Recognition Method Based on Image Color and Texture Features, Comp. Math. Methods Med. **2018** (2018) 10 pages.

[39] I. Kononenko, Machine learning for medical diagnosis: history, state of the art and perspective. Artif. Intell. Med. 23 (2001) 89–109, https://doi.org/10.1016/S0933-3657(01)00077-X.

[40] Y.-J. Hu, T.-H. Ku, R.-H. Jan, K. Wang, Y.-C. Tseng, S.-F. Yang, Decision tree-based learning to predict patient controlled analgesia consumption and readjustment. BMC Med. Inform. Decis. Mak. 12 (2012) 131, https://doi.org/10.1186/1472-6947-12-131.

[41] R. Schapire, Explaining AdaBoost. (2013), https://doi.org/10.1007/978-3-642-41136-6_5.

[42] https://medium.com/the-theory-of-everything/understanding-activation-functions-in-neural-networks-9491262884e0.

[43] P. Samui, S.S. Roy, V.E. Balas (Eds.), Handbook of Neural Computation, Academic Press, 2017.

[44] V.E. Balas, S.S. Roy, D. Sharma, P. Samui (Eds.), Handbook of Deep Learning Applications, In: vol. 136, Springer, 2019.

[45] A. Bose, S.S. Roy, V.E. Balas, P. Samui, Deep learning for brain computer interfaces, in: Handbook of Deep Learning Applications, Springer, Cham, 2019, pp. 333–344.

10

Driver drowsiness detection using heart rate and behavior methods: A study

Anmol Wadhwa and Sanjiban Sekhar Roy

School of Computing Science and Engineering, Vellore Institute of Technology, Vellore, Tamil Nadu, India

1   Introduction

There has been exponential growth in the automobile manufacturing industry from 1950 to 2019 [1]. Estimated global production of automobiles in 2018 was 95,634,593 cars. The Indian automobile industry is the fourth most prominent automobile manufacturing industry in the world with annual sales growth up to 9.5% per year and production of 4.02 million units in 2017 [1]. Automobile sales are expected to reach $251.4–282.8 billion by 2026 [2]. Statistics show that automobile sales have been trending up and simultaneously increasing employment. This sector has increased social ties among various countries, infrastructure and the circulation of imports and exports. Automobile has made it easy to travel to distant or isolated places. Steve Magge, a leading expert on human health and radiation, once said, "Most people do not realize that they are one car crash away from death" [1]. The World health Organization's (WHO) global status report on violence and injury prevention (2018) reported 1.35 million annual road traffic deaths. Motor vehicle accidents are the primary killer of individuals aged 5–29 years old. Road deaths in India totaled 0.149 million in 2018 with Uttar Pradesh registering a maximum spike in the number of deaths.

Driving is a cognitively complex task that involves most parts of the human body. As such, a driver's medical state plays an important role in coordination and response time during driving.

Taking into account the risks posed by a driver's medical health, researchers have created different methodologies to recognize driver drowsiness. In the last 10 years, numerous algorithms have been developed for determining driver drowsiness. These techniques include recording driver behavior [3, 4], measuring driver brain signals [5], and assessing performance of vehicles [6].

One popular technique is the bio-signal approach, which is considered the most important technology for identifying driver drowsiness. Different types of datasets exist for measuring electrical signals in the brain, including electroencephalography (EEG), electrocardiography (ECG), electromyography (EMG), eye development and electrooculography (EOG), and functional near-infrared spectroscopy (fNIRS) for measuring blood flow in the brain [7]. A brain-computer interface (BCI) or neural control interface is used for controlling, monitoring, and communicating with the brain using electrical signals. Research on BCI can be characterized into three fundamental classes: reactive, active, and inactive brain activities [8–10]. In contrast with other brain imaging modalities, fNIRS is low cost, versatile, has greater temporal resolution, and is easier to use. In addition, fNIRS [9, 11–14] is not vulnerable to electrical noise because it utilizes optical signals for detection of brain tasks. Most researchers have used the following methods for analyzing brain signals: relative power level (RPL), mean of oxyhemoglobin (HbO), peak of signal, addition of maximum values in signal, and RR peak from the ECG record as RR maximum values for measuring normal and unstable heartbeat time. In any case, manual feature extraction is required for these algorithms [15–17, 18-28]. Recent methods have used technologies that automatically characterize driver features between sleepy and non-lazy states. These include [10, 29, 30, 31-51] classifiers such as k-nearest neighbors (KNNs), linear regression, and support vector machine (SVM). Feature selection and feature extraction are the two fundamental steps required when using a machine learning algorithm [30]. The deep learning approach is becoming more popular because of automatic feature extraction and selection. Deep learning models not only extract these features, but also learn from them. Convolutional neural networks (CNNs) are generally used in image classification and segmentation [52–54, 55-57].

In this chapter we review two methodologies. The first is driver drowsiness detection using heartbeat with the help of Arduino. The second uses behavior methods like yawning and eye movement to detect driver drowsiness.

2 Heartbeat detection method

Monitoring heartbeat, temperature, and blood pressure are the first steps in determining whether a person is healthy or sick. These techniques have been part of any medical checkup for decades. The only change now is the technologies being used. Many researchers have proposed systems for measuring heart rate with only pulse detection. The following sections describe them in short.

2.1 Components required

2.1.1 Grove ear clip

The heart rate ear clip has a receiver module and sensor used to calculate heart rate. The ear stick is worn on the ear flap or at the tip of the finger, and it has a heart rate sensor embedded. It works on low power and is highly sensitive and user friendly [58, 59]. It is connected to an analog-to-digital converter (ADC), so the converted signal can work as an input to the yellow wire of the Arduino system.

2.1.2 Arduino UNO

The Arduino UNO is a user-friendly device derived from the ATmega328. There are 14 digital pins and 6 analog pins. It also has a ceramic resonator (16 MHz), a USB port, a power jack, and a reset button [59]. It has a USB cable that connects to a laptop/desktop computer as the power source. Alternatively, it can be powered using a battery. Fig. 1 shows the Arduino UNO board we used in our experiment.

2.2 Circuit diagram

The Arduino is connected to a power supply of 5 V, to a buzzer via a digital pin and ground pin, and to the ear grove sensors. After connecting the components of the circuit, the ear clip heart sensor is worn by the driver on the fingertip or ear flap, which generates the driver's pulse, which is then given as an input to the Arduino. The detailed connection of the circuit is described in Figs. 2 and 3. The circuit processing speed mainly depends on the Arduino microcontroller, which has a 12-bit ADC and 16-MHz process execution speed.

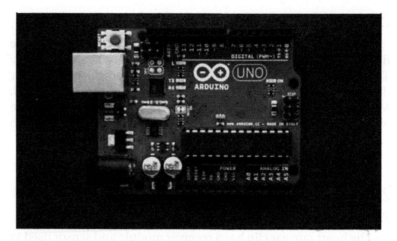

Fig. 1 Arduino UNO. Photo by Harrison Broadbent on Unsplash.

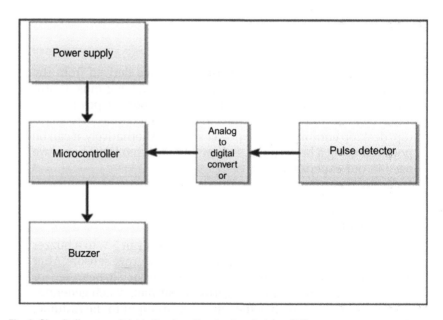

Fig. 2 Circuit diagram of detecting heartbeat using Arduino [58].

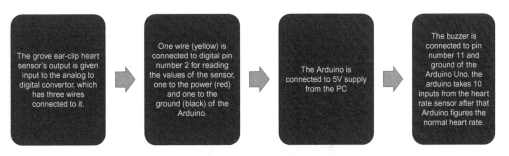

Fig. 3 Circuits diagram connection.

2.2.1 A graphical representation of driver drowsiness detection

After the car starts, the Arduino is triggered and begins measuring the driver's hear rate. If the driver's heart rate crosses a certain threshold value, the system triggers a buzzer. The flow diagram of the whole process is depicted in Fig. 4. A sample heartbeat signal

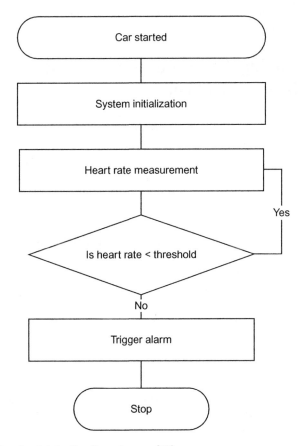

Fig. 4 Heartbeat detection flow diagram [58].

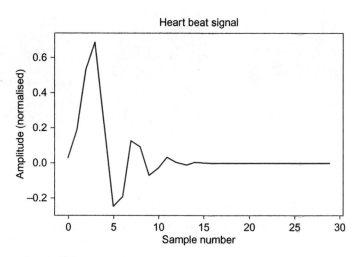

Fig. 5 Detected heart rate [58].

with amplitude on the y axis and number of beats on the x axis is shown in Fig. 5.

3 Detecting driver drowsiness using behavioral method

Behavioral strategies measure different levels of drowsiness among drivers using a camera fixed near the steering wheel for observing the different parts of face to assess the state of eyelids, the movement of the head, and the rate of eye flickering and yawning. After obtaining these features from the camera feed, the proposed algorithm is applied to determine the degree of the drowsiness, commonly by artificial intelligence (AI) systems. These models first extract features using models like SVMs and CNNs, then classifies these extracted features by long short term memory networks (LSTM) as either drowsy or not drowsy. The most difficult part of this method is finding the appropriate dataset based on specific race and not violating the integrity or confidentiality of the driver. Fig. 6 shows a typical structure of the algorithm that is most utilized for detecting driver drowsiness. Following are different characteristics used by various algorithms for classification.

3.1.1 Shut eye examination:

The state of the eye is a significant component that generally decides driver drowsiness. To determine the state of the eyes, shut eye examination (SEE) [60] and eye angle ratio (EAR) can be used.

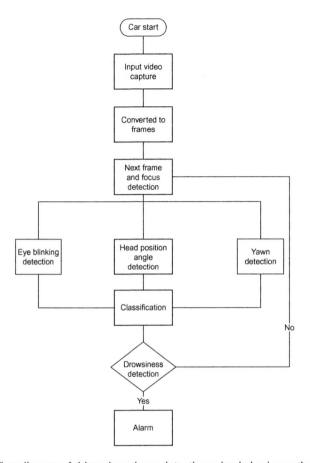

Fig. 6 Flow diagram of driver drowsiness detection using behavior method.

EAR, as presented by Cech and Soukupova [60, 14], is defined as length-to-width ratio of the eye. SEE is defined as eye shut percentage over some time period. EAR categorizes the person as sleepy or not sleepy, whereas SEE categorizes eyes as closed or open.

3.1.2 Eye flicker rate

This eye flicker rate approach calculates recurrence of flickering. The typical flickering rate of the eye every moment is around 10 which signifies that the driver is tired and flickering rate has been diminished.

3.1.3 Evaluation of yawning

Yawning be triggered by fatigue. A driver who is yawning may potentially nod off. A handful of techniques have been proposed to identify enlargement of the mouth and other yawning attributes, for example, the location of the upper and lower lips [14, 63].

3.1.4 Algorithm for predicting drowsiness

Following are the steps for detecting driver drowsiness:
- Video recording
  First a camera is placed near the steering wheel to record video of the driver. Then, the video is converted into frames.
- Face detection
  Second, the driver's face is recognized from the video using an algorithm by Viola and Jones [14, 63]. Then, the next task is to extract the characteristics of a drowsy face.
- Extraction of features
  If the driver's face is discovered properly, its features can easily be extracted by using various techniques, for example, localization of landmark, histogram of sloping gradients (HOG), and local binary patterns (LBPs). Extracted features are then prepared further to measure SEE or EAR for eye examination or mouth-based techniques for yawning discovery [61].
- Grouping
  This stage comprises classifiers that are used to classify driver drowsiness. In this stage, the classifier identifies attributes of drowsiness dependent on weighted parameters. If the accuracy of classification compared to a given test set is greater than 90%, then classification is deemed successful. If not, changes in training parameters need to be made.

Behavioral strategies have different constraints on the grounds that they can be influenced by lighting conditions, camera arrangements, and the frames per second used to catch pictures of the driver's face. Lighting changes can usually be exterminated by making use of infrared (IR) cameras. Fig. 6 presents a flow diagram of the preceding steps.

3.2 Analysis of various classifications for driver drowsiness detection

Different measures have been utilized in various studies for recognizing a face and then separating out its features. It is important to note that some studies used contrasting datasets that support only the researchers' personal techniques. This is due to the absence of institutionalized datasets that could be used as a standard for classification. Therefore it is difficult to think about methodologies by just assessing precision. In the following section we discuss various models that have been used in previous research.

3.2.1 Support vector machines

SVMs are part of supervised machine learning algorithms that are mainly used for analyzing data required in classification and regression analysis. SVM models find the hyperplane that best classifies features into different domains. Points near to the hyperplane are known as support vector points and the vectors' distance from the hyperplane are called margins. The greater the length of margins the more probability there is of correctly classifying the point [62, 63]. Many authors have applied SVM models for different datasets in their research. The writers of [64] proposed a framework used for detecting the eyes and face. The training set in the framework contains states like open eye, open mouth, and so on. After testing, the accuracy of this model was found to be 100%, however, this result is achievable at slower frame rates, which leads us to miss facial expressions.

3.2.2 Hidden Markov model

Hidden Markov models (HMMs) are statistical models based on augmenting Markov chains. Markov chains are defined by a graph that experiences transition from one state to another according to rules of probability.

Prediction in Markov chains is always dependent on the current state, not on any previous states. HMMs uses a hidden Markov chain, and each state in this chain randomly generates one out of k observations. They were developed in the 1960s and 1970s by Leonard Baum [65] and are used in applications such as facial expression detection, computer virus classification, and gene annotation [66, 67]. The authors of [66] suggested a new facial feature by identifying the change in wrinkles on the face. They used an IR camera that can be used in both day and night conditions. The main drawback of this framework is that it can yield false outcomes when aged individuals are involved in the experiment. In [67] HMM was used to track eye movement based on color and geometric features of the eye, but this model failed to recognize eye movement if the driver was not looking forward and if the experiment was performed during nighttime conditions.

3.2.3 Convolutional neural network

CNNs are a class of neural network that allow greater extraction of features from captured images [68]. Unlike classical models, CNNs take image data, train the model, and then classify the features automatically for healthier classification. There are three principal components of CNNs: convolution, maxpooling, and activation function. CNNs are used in many applications like image recognition, face recognition, and video analysis [68]. Yann Le Cun [69] was the first researcher to use CNN in computer vision applications, but results were not good until he used deep CNN for object recognition. In [68] driver drowsiness detection was accomplished using representation learning that used Viola and Jones' algorithm for face classification. In this experiment, the pictures were cropped to 48 × 48 frames and then passed to the outmost layer of the framework, which comprised 20 filters. Then, its output was fed to the softmax layer. This system doesn't consider face position. Other researchers have used a 3D deep neural network in which the face is passed through a combination of two more filters [70], thus making the system applicable even when the driver changes head position.

3.2.4 Comparative analysis of classifiers for driver drowsiness detection

Each classifier has its own pros and cons. It is not necessary that each classifier be suitable for different datasets with specific features. HMMs use features like probabilities, eye state, hidden Markov state, and face position for classification. SVMs use parameters like yawning, eye state, eye closure, and so on for classification. CNNs use parameters like eye state and facial features for classification. Selection of the appropriate model is dependent on system requirements, accuracy, speed of classification, and error rate. Error rates for HMMs, SVMs, and CNNs are 4% [71], 4.7% [72], and 9%, respectively, according to the literature [70]. HMMs can capture dependencies in the measurement and detect variance of the device's power, but HMMs are limited by their layer intrinsic structure. CNNs provide very good accuracy in image segmentation, but the process is slow and requires a large training dataset. These methods are not suitable for large datasets and are not very effective for missing data and noisy data.

4 Conclusion

In this chapter we discussed recent popular techniques for driver drowsiness detection. The first is the heartbeat method and the second is the behavior method. For the first technique, we briefly described the Arduino system, ear glove sensor, and circuit diagram. For the second technique, we reviewed different machine learning algorithms. The main objective of using machine learning is to detect slight changes in the appearance of the driver's facial expressions in the eye or mouth region so that drowsiness can be captured quickly. SVMs, CNNs, and HMMs are the three main techniques used for this purpose. Existing methods do not require any vehicle adjustment or circuitry for measuring driver heart rate or brain signals, however, these algorithms are really slow and are not reliable as compared to techniques such as ECG.

References

[1] D. Mohan, O. Tsimhoni, M. Sivak, M.J. Flannagan, Road Safety in India: Challenges and Opportunities, UMTRI-2009-1The University of Michigan Transportation Research Institute, Ann Arbor, MI, 2018, pp. 1–57.

[2] Expert Committee on Auto Fuel Policy, Urban Road Traffic and Air Pollution in Major Cities, vol. 1, Government of India, New Delhi, 2002, pp. 1–395.

[3] K. Fujiwara, E. Abe, K. Kamata, C. Nakayama, Y. Suzuki, T. Yamakawa, T. Hiraoka, M. Kano, Y. Sumi, F. Masuda, M. Matsuo, H. Kadotani, Heart rate variability-based driver drowsiness detection and its validation with EEG, IEEE Trans. Biomed. Eng. 66 (6) (2019) 1769–1778.

[4] S.K.L. Lal, A. Craig, A critical review of the psychophysiology of driver fatigue, Biol. Psychol. 55 (3) (2001) 173–194.

[5] J. Vicente, P. Laguna, A. Bartra, R. Bailon, Drowsiness detection using heart rate variability, Med. Biol. Eng. Comput. **54** (6) (2010) 927–937.

[6] T. Nguyen, S. Ahn, H. Jang, S.C. Jun, J.G. Kim, Utilization of a combined EEG/NIRS system to predict driver drowsiness, Sci. Rep. 7 (2017) 43933.

[7] D.A. Boas, C.E. Elwell, M. Ferrari, G. Taga, Twenty years of functional near-infrared spectroscopy: introduction for the special issue, NeuroImage 85 (2014) 1–5.

[8] N.K. Logothetis, What we can do and what we cannot do with fMRI, Nature 453 (7197) (2008) 869–878.

[9] M. Strait, M. Scheutz, What we can and cannot (yet) do with functional near infrared spectroscopy, Front. Neurosci. 8 (2014) 117.

[10] BoseA., RoyS.S., BalasV.E., SamuiP., Deep learning for brain computer interfaces, in: Handbook of Deep Learning Applications, Springer, Cham, 2019, , pp. 333–344.

[11] M.P. van den Heuvel, H.E.H. Pol, Exploring the brain network: a review on resting-state fMRI functional connectivity, Eur. Neuropsychopharmacol. 20 (8) (2010) 519–534.

[12] M. Welvaert, Y. Rosseel, A review of fMRI simulation studies, PLoS One 9 (7) (2014) e101953.

[13] S. Coyle, T. Ward, C. Markham, Brain–computer interfaces: a review, Interdiscip. Sci. Rev. 28 (2) (2003) 112–118.

[14] B.E. Boser, I.M. Guyon, V.N. Vapnik, A training algorithm for optimal margin classifiers, in: Proc. fifth Annu. Work. Comput. Learn. theory, 1992, pp. 144–152.

[15] D. Wu, J.-T. King, C.-H. Chuang, C.-T. Lin, T.-P. Jung, Spatial filtering for EEG-based regression problems in brain–computer interface (BCI), IEEE Trans. Fuzzy Syst. 26 (2) (2018) 771–781.

[16] S. Barua, M.U. Ahmed, C. Ahlström, S. Begum, Automatic driver sleepiness detection using EEG, EOG and contextual information, Expert Syst. Appl. 115 (2019) 121–135.

[17] X. Zhang, J. Li, Y. Liu, Z. Zhang, Z. Wang, D. Luo, X. Zhou, M. Zhu, W. Salman, G. Hu, C. Wang, Design of a fatigue detection system for high-speed trains based on driver vigilance using a wireless wearable EEG, Sensors 17 (3) (2017) 486.

[18] K. Ismail, T. Sayed, N. Saunier, C. Lim, Automated analysis of pedestrian-vehicle conflicts using video data, Transp. Res. Rec. J. Transp. Res. Board 2140 (1) (2009) 44–54.

[19] P. Philip, T. Åkerstedt, Transport and industrial safety, how are they affected by sleepiness and sleep restriction? Sleep Med. Rev. 10 (5) (2006) 347–356.

[20] C.F.P. George, Sleep apnea, alertness, and motor vehicle crashes, Am. J. Respir. Crit. Care Med. 176 (10) (2007) 954–956.

[21] J.A. Owens, T. Dearth-Wesley, A.N. Herman, R.C. Whitaker, Drowsy driving, sleep duration, and chronotype in adolescents, J. Pediatr. 205 (2019) 224–229.

[22] C.C. Liu, S.G. Hosking, M.G. Lenné, Predicting driver drowsiness using vehicle measures: recent insights and future challenges, J. Saf. Res. 40 (2009) 239–245.

[23] T.O. Zander, C. Kothe, Towards passive brain–computer interfaces: applying brain–computer interface technology to human–machine systems in general, J. Neural Eng. 8 (2) (2011) 025005.

[24] M. Ferrari, V. Quaresima, A brief review on the history of human functional near-infrared spectroscopy (fNIRS) development and fields of application, NeuroImage 63 (2) (2012) 921–935.

[25] N. Naseer, K.-S. Hong, fNIRS-based brain-computer interfaces: a review, Front. Hum. Neurosci. 9 (2015) 3.

[26] L.F. Nicolas-Alonso, J. Gomez-Gil, Brain computer interfaces, a review, Sensors 12 (2) (2012) 1211–1279.

[27] C.-H. Chuang, L.-W. Ko, Y.-P. Lin, Y.-P. Jung, C.-T. Lin, Independent component ensemble of EEG for brain–computer interface, IEEE Trans. Neural Syst. Rehabil. Eng. 22 (2) (2014) 230.

[28] M.J. Khan, K.-S. Hong, Passive BCI based on drowsiness detection: an fNIRS study, Biomed. Opt. Express 6 (10) (2015) 4063–4078.

[29] A. Bandhu, S.S. Roy, Classifying Multi-Category Images Using Deep Learning: A Convolutional Neural Network Model.

[30] Y. Liu, H. Ayaz, Speech recognition via fNIRS based brain signals, Front. Neurosci. 12 (2018) 695.

[31] S. Ahn, T. Nguyen, H. Jang, J.G. Kim, S.C. Jun, 'Exploring neurophysiological correlates of drivers' mental fatigue caused by sleep deprivation using simultaneous EEG, ECG, and fNIRS data, 'T. Wilcox and M. Biondi,' "fNIRS in the Developmental Sciences" Wiley Interdiscip. Rev. Cogn. Sci. 6 (3) (2015) 263–283.

[32] K.-S. Hong, N. Naseer, Y.-H. Kim, Classification of prefrontal and motor cortex signals for three-class fNIRS–BCI, Neurosci. Lett. 587 (2015) 87–92 L.C. Schudlo, T. Chau, "Development and testing an online near infrared spectroscopy brain– computer interface tailored to an individual with severe congenital motor impairments.".
[33] L.C. Schudlo, T. Chau, Development of a ternary near-infrared spectroscopy brain-computer interface: Online classification of verbal fluency task, stroop task and rest, Int. J. Neural Syst. 28 (4) (2018) 1750052.
[34] T. Gateau, H. Ayaz, F. Dehais, In silico vs. over the clouds: on-the-fly mental state estimation of aircraft pilots, using a functional near infrared spectroscopy based passive-BCI, Front. Hum. Neurosci. 12 (2018) 187.
[35] K.J. Verdière, R.N. Roy, F. Dehais, Detecting pilot's engagement using fNIRS connectivity features in an automated vs. manual landing scenario, Front. Hum. Neurosci. 12 (2018) 6.
[36] S.S. Roy, R. Sikaria, A. Susan, A Deep Learning Based CNN Approach on MRI for Alzheimer's Disease Detection, Intelligent Decision Technologies, 2019.
[37] D. Mittal, D. Gaurav, S.S. Roy, An effective hybridized classifier for breast cancer diagnosis, in: IEEE International Conference on Advanced 2015, 2015.
[38] P. Samui, S.S. Roy, V.E. Balas, Handbook of Neural Computation, (2017).
[39] D. Pei, M. Burns, R. Chandramouli, R. Vinjamuri, Decoding asynchronous reaching in electroencephalography using stacked auto encoders, IEEE Access 6 (2018) 52889–52898.
[40] F. Nasoz, O. Ozyer, C.L. Lisetti, N. Finkelstein, Multimodal affective driver interfaces for future cars, in: Proc. ACM Int. Multimedia Conf. Exhibition, 2002, pp. 319–322.
[41] Solarbotics, https://solarbotics.com/product/29170/, 2019.
[42] M. Asaduzzaman Miah, M.H. Kabir, M.S.R. Tanveer, M.A.H. Akhand, Continuous heart rate and body temperature monitoring system using Arduino UNO and Android device, in: 2015 2nd International Conference on Electrical Information and Communication Technologies (EICT), Khulna, 2015, pp. 183–188.
[43] M.S. Bin Zainal, I. Khan, H. Abdullah, Efficient drowsiness detection by facial features monitoring, Res. J. Appl. Sci. Eng. Technol. 7 (11) (2014) 2376–2380.
[44] T. Nakamura, A. Maejima, S. Morishima, Detection of driver's drowsy facial expression, in: Proc.—2nd IAPR Asian Conf. Pattern Recognition, ACPR 2013, 2013, pp. 749–753.
[45] P. Viola, M.J. Jones, Robust real-time face detection, Int. J. Comput. Vis. 57 (2) (2004) 137–154.
[46] A. Ben-Hur, J. Weston, A user's guide to support vector machines, Methods Mol. Biol. 609 (2010) 223–239.
[47] T.S. Prasad, N.R. Kisore, Application of hidden Markov model for classifying metamorphic virus, in: Souvenir 2015 IEEE Int. Adv. Comput. Conf. IACC 2015, 2015, pp. 1201–1206.
[48] B.-J. Yoon, Hidden Markov models and their applications in biological sequence analysis, Curr. Genomics 10 (6) (2009) 402–415.
[49] U. Karn, An Intuitive Explanation of Convolutional Neural Networks, [Online]. Available: https://ujjwalkarn.me/2016/08/11/intuitive-explanation-convnets/, 2016. Accessed 9 June 2017.
[50] A. Krizhevsky, I. Sutskever, G.E. Hinton, ImageNet classification with deep convolutional neural networks, Adv. Neural Inf. Process. Syst. (2012) 1–9.
[51] M. Ngxande, J. Tapamo, M. Burke, Driver drowsiness detection using behavioral measures and machine learning techniques: a review of state-of-art techniques, in: 2017 Pattern Recognition Association of South Africa and Robotics and Mechatronics (PRASA-RobMech), Bloemfontein, 2017, pp. 156–161.

[52] M. Zhang, M. Gong, Y. Mao, J. Li, Y. Wu, Unsupervised feature extraction in hyperspectral images based on wasserstein generative adversarial network, IEEE Trans. Geosci. Remote Sens. 57 (5) (2019) 2669–2688.

[53] Y. Yang, Z. Ye, Y. Su, Q. Zhao, X. Li, D. Ouyang, Deep learning for in vitro prediction of pharmaceutical formulations, Acta Pharm. Sin. B 9 (1) (2019) 177–185.

[54] M. Usama, B. Ahmad, J. Wan, M.S. Hossain, M.F. Alhamid, M.A. Hossain, Deep feature learning for disease risk assessment based on convolutional neural network with intra-layer recurrent connection by using hospital big data, IEEE Access 6 (2018) 67927–67939.

[55] S.S. Roy, C. Pratyush, C. Barna, Predicting Ozone Layer Concentration Using Multivariate Adaptive Regression Splines, Random Forest and Classification and Regression Tree, (2018).

[56] S.S. Roy, R. Sikaria, A. Susan, A Deep Learning Based CNN Approach on MRI for Alzheimer's Disease Detection, (2019) pp. 495–505.

[57] R. Biswas, A. Vasan, S.S. Roy, Dilated deep neural network for segmentation of retinal blood vessels in fundus images. Iran. J. Sci. Technol. Trans. Electr. Eng. 44 (2020) 505–518, https://doi.org/10.1007/s40998-019-00213-7.

[58] S.F. Barrett, Arduino microcontroller processing for everyone!, in: Estimating Heating Load in Buildings Using Multivariate Adaptive Regression Splines, Extreme Learning Machine, a Hybrid Model of MARS and ELM, Morgan and Claypool Publishers, 2010.

[59] I.H. Choi, C.H. Jeong, Y.G. Kim, Tracking a driver's face against extreme head poses and inference of drowsiness using a hidden Markov model, Appl. Sci. 6 (5) (2016).

[60] X. Zhang, J. Li, Y. Liu, Z. Zhang, Z. Wang, D. Luo, X. Zhou, M. Zhu, W. Salman, G. Hu, C. Wang, Design of a fatigue detection system for high-speed trains based on driver vigilance using a wireless wearable EEG, Sensors 17 (3) (2017) Art. No.: 486.

[61] S. Lemm, B. Blankertz, T. Dickhaus, K.-R. Müller, Introduction to machine learning for brain imaging, NeuroImage 56 (2) (2011) 387–399.

[62] M. Sabet, R.A. Zoroofi, K. Sadeghniiat-Haghighi, M. Sabbaghian, A new system for driver drowsiness and distraction detection, in: ICEE 2012—20th Iran. Conf. Electr. Eng, 2012, pp. 1247–1251.

[63] S.S. Roy, A. Gupta, A. Sinha, R. Ramesh, Cancer data investigation using variable precision Rough set with flexible classification, in: Proceedings of the Second International Conference, 2012.

[64] CS231n Convolutional Neural Networks for Visual Recognition. [Online]. Available: http://cs231n.github.io/convolutionalnetworks/. Accessed 9 June 2017.

[65] R. Zimmermann, L. Marchal-Crespo, J. Edelmann, O. Lambercy, M.-C. Fluet, R. Riener, M. Wolf, R. Gassert, Detection of motor execution using a hybrid fNIRS-biosignal BCI: a feasibility study, J. Neuroeng. Rehabil. (2013) M. Rea, M. Rana, N. Lugato, P. Terekhin, L. Gizzi, D. Brötz, and A. Caria, "Lower limb movement preparation in chronic stroke: A pilot study toward an fNIRS-BCI for gait rehabilitation.

[66] AmariS., The Handbook of Brain Theory and Neural Networks, MIT Press, 2003.

[67] S.S. Roy, A. Mallik, R. Gulati, M.S. Obaidat, P.V. Krishna, A Deep Learning Based Artificial Neural Network Approach for Intrusion Detection, (2011).

[68] C. Chakraborty, S.S. Roy, M.J. Hsu, G. Agoramoorthy, Landscape mapping of functional proteins in insulin signal transduction and insulin resistance: a network-based protein-protein interaction analysis, PLoS One 6 (1) (2011) 1–7 e16388.

[69] DwivediK., BiswaranjanK., SethiA., Drowsy driver detection using representation learning, 2014 IEEE International Advance Computing Conference (IACC), IEEE, 2014, , pp. 995–999.

[70] S.S. Roy, P. Samui, I. Nagtode, et al., Forecasting heating and cooling loads of buildings: a comparative performance analysis. J. Ambient Intell. Human. Comput. 11 (2020) 1253–1264, https://doi.org/10.1007/s12652-019-01317-y.

[71] G.J. Al-Anizy, M.J. Nordin, M.M. Razooq, Automatic driver drowsiness detection using Harr algorithm and support vector machine techniques, Asian J. Appl. Sci. 8 (2) (2015) 149–157.

[72] RoyS.S., RoyR., BalasV.E., Estimating heating load in buildings using multivariate adaptive regression splines, extreme learning machine, a hybrid model of MARS and ELM, Renew. Sustain. Energy Rev. **82** (2018) 4256–4268.

Innovative classification, regression model for predicting various diseases

B.K. Tripathy, M Parimala, and G Thippa Reddy
School of Information Technology and Engineering, VIT Vellore, Vellore, Tamil Nadu, India.

1 Introduction

Data mining plays a major role in the healthcare industry. It can be used to enhance healthcare processes systematically and determine the best healthcare techniques at the lowest cost. Data mining performs analysis of large datasets to discover hidden patterns that can help to forecast or predict future events. Applications of data mining are vast, including in the retail industry, telecom services, automotive industry, and life sciences [1]. However, data mining in health care [2] is the need of the hour. Data can be generally categorized as either descriptive analytics, which describe what happened; predictive analytics, which denote events occurring in the future; and prescriptive analytics, which determine solutions for certain issues. The latter involves mining hidden patterns to build predictive models.

There are various data mining techniques, such as association rule generation, classification, clustering, and outlier analysis. An association rule is expressed as X → Y, where X and Y are sets of items in a transactional database. The rules generated have support and confidence levels greater than or equal to the minimum threshold value. This technique is widely used in market basket analysis for finding frequent items purchased together. For example, a customer who buys bread also purchases jam and milk. Apriori is a popularly used association rule mining algorithm. Classification [3] is a supervised learning technique that depends on class labels or target variables. The training data set is analyzed and a model is constructed based on target variables that aims to categorize new unlabeled records. The decision tree is one of the widely used classification algorithms. Outlier detection is a method for identifying noisy data from the given dataset. The outlier detection algorithm can be applied in analyzing customer churn, forecasting unusual weather conditions, and detecting fraudulent customers.

Some of the soft computing techniques such as neural networks, genetic algorithms, rough set techniques, and support vector machines are also used with data mining techniques for optimizing results, to search for dominating attributes in a given dataset and to handle high dimensional data. Based on the nature of the dataset, data mining can be categorized as sequence extraction, web mining, text mining, or spatial data mining. Data mining techniques are applied in the field of business and e-commerce such as stock market analysis and financial prediction. Other applications are in the fields of scientific engineering and healthcare data, like finding relationships between genomic data, predicting patterns in sensor data, intensive computing in simulated data, diagnosing disease in infected people, and predicting disease outbreak. They are in use for extracting meaningful information from voluminous multimedia data in fields like banking, customer retention, targeted marketing, and crime detection.

The first part of this chapter discusses the various classification models and methods used for classifying diseases. The second part explains the different regression models designed for specific types of diseases like diabetes, cardiac disease, and epidemiological disease.

2 Classification

Machine learning can be either supervised or unsupervised. In supervised learning the machine learning algorithm will know in advance what will be the target/class label (i.e., the algorithm will be fed/informed about the value that it has to predict). In unsupervised learning, the machine learning algorithm will not be aware of the target/class label (i.e., the algorithm will not have information about what value it has to predict).

Classification is one of the most popular supervised learning techniques used in data mining/machine learning [4–7]. The classification algorithms will learn about the characteristics/dependencies existing between several fields with the help of training data. The classification algorithm learns more about the data if the quantity of data is more. The classification algorithm will take the values of several fields as input and will predict the value for class label.

Classification identifies to which category a new observation belongs to. For example, given a set of risk factors like age, blood sugar levels, blood pressure, cholesterol, and so on of a person, a classification algorithm can categorize whether the person is a potential heart disease patient or not. We can also build a classification model for determine whether an application for a loan is going to be risky or safe.

Classification can be applied in many fields. Some of the applications where classification can be applied include health care, weather forecasting [5], image processing, marketing and retail, manufacturing, telecommunications, intrusion detection, education, fraud detection, and many others.

Some of the popular data mining algorithms include decision trees, Bayesian classifiers, neural networks, k-nearest neighbors, support vector machines, linear regression, logistic regression, and so on.

2.1 Steps in developing a classifier

Any classification algorithm involves the following steps:
(1) Building the classifier or model
(2) Using the classifier for classification

2.2 Building the classifier or model

In this phase, the classification algorithm or classifier will learn about the patterns, dependencies, or relationships between several attributes. This phase can be called the learning phase. In this step, classification algorithms are used to build the classifier. Classification algorithms use the dataset tuples and their associate class labels from the training dataset to build the classifier. Each tuple that constitutes the training set is referred to as a category or class. The outcome of this phase is the classification rules as shown in Fig. 1, which shows an example for building the classifier. In this example the training datasets include several fields like name, age, and income of several persons who have applied for a loan, and the class label is loan_decision, which can be either risky or safe. The classification algorithm will use this dataset to train itself and come up with some classification rules like, if age = youth then loan_decision = risky, if income = high then loan_decision = safe, and so on.

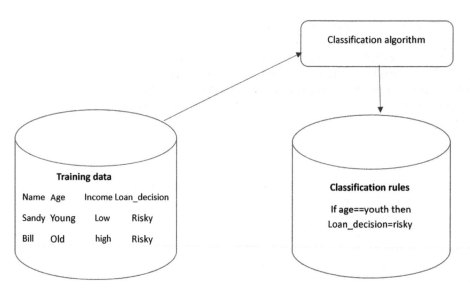

Fig. 1 Example for building the classifier [7].

2.3 Using classifier for classification

In this step, the classification model that was built in the previous step classifies the data. The classifier built uses the test data to determine how accurate the classification rules are. In this way we can decide whether classification rules built by the classifier can be used on new data tuples if the accuracy is good enough.

Fig. 2 shows an example for usage of the classifier built in the previous step to classify a testing tuple. In this example, a new tuple whose name is John, whose age is middle-aged, and whose income is low, predicts that class label loan_decision is "risky."

2.4 Classification issues

Before training a classifier several things have to be taken care of. Some of these issues are:

Data cleaning: The data collected may have several missing data. Also some of the data may be noisy. The classifier may not give accurate results if the dataset has these properties. Hence they have to be traded before the dataset is fed to the classifier for training purposes.

Relevance analysis: The dataset may have several attributes that are not relevant for the task. Also some of the attributes may be redundant. For example, if the classifier has to predict

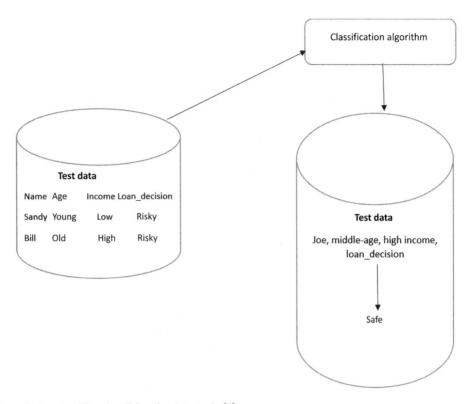

Fig. 2 Example for classifier classifying the data tuple [7].

whether a person will be buying a product or not, then the attribute "name" is not useful. In the same way, if the dataset has two attributes, date of birth and age, they are redundant. These kind of attributes have to be taken care of before building the classifier as their presence will increase the complexity of the classifier.

Data transformation and reduction: In some cases, the dataset may be collected from several sources. In those cases, every data source might be following their own notations in storing the data. For example, consider the attribute total sales. If the data source is in India, the units in which the data stores will be in rupees, and if the data source is in the United Kingdom, then the units will be in pounds, and so on. Hence while integrating the data, the data has to be brought into a common format so that it will be easy to process. Also, in some cases we may want to scale down the data and fit the data within some range. For example, assume that for the attribute salary, the range in the given dataset is ₹10,000–12,00,000. We may want to fit the values of this attribute within a scale of 0–1. In this case we may have to use normalization techniques to fit the data within a given range. In some cases, we may want to reduce the size of the dataset for optimizing storage. For this purpose we have to use several data reduction techniques.

2.5 Testing a classification model

We can test the built classifier by applying it with some test data whose value for the class label is known to us. Normally the test data will come from the same dataset that is collected for classification. Some percentage of the data will be used for training and the remaining percentage of the data will be used for testing the classifier. We will be using some metrics for testing the effectiveness of the classifier. If the performance of the classifier in those metrics is good, we can use the model on new data for future predictions. Some of the important and widely used metrics are:

2.5.1 Accuracy

This is the percentage of correct predictions that a classifier has made when compared to the actual value of the label in the testing phase. Accuracy can be calculated as:

$$\text{Accuracy} = (TN + TP)/(TN + TP + FN + FP)$$
$$= (\text{Number of correct assessments})/\text{Number of all assessments})$$

where TP is true positives, TN is false negatives, FP is false positives, and FN is false negatives.

If the class label of a record in a dataset is positive, and the classifier predicts the class label for that record as positive, then it is called as true positive. If the class label of a record in a dataset is negative, and the classifier predicts the class label for that record as negative, then it is called as true negative. If the class label of a record in a dataset is positive, but the classifier predicts the class label for that record as negative, then it is called as false negative. If the class label of a record in a dataset is negative, but the classifier predicts the class label for that record as positive, then it is called as false positive.

2.5.2 Sensitivity

This is the percentage of true positives that are correctly identified by the classifier during testing. It is calculated as:

$$TP/(TP + FN) = (\text{Number of true positive assessment})/(\text{Number of all positive assessment})$$

2.5.3 Specificity

This is the percentage of true negatives that are correctly identified by the classifier during testing. It is calculated as:

$$TN/(TN+FP) = (\text{Number of true negative assessment})/(\text{Number of all negative assessment})$$

2.5.4 Confusion matrix

Using a confusion matrix, we can say total number of times a classifier predicted correctly and incorrectly in comparison with the actual values of the labels during the testing phase. It is n-by-n, where n is the number of classes.

In the example given in Fig. 3, the model predicted correctly the positive class for affinity_card 516 times, and predicted it incorrectly 25 times. The model correctly predicted the negative class for affinity_card 725 times and predicted it incorrectly 10 times. The following can be computed from this confusion matrix:

- The model made 1241 correct predictions (516 + 725)
- The model made 35 incorrect predictions (25 + 10)
- There are 1276 total scored cases (516 + 25 + 10 + 725)
- The error rate is 35/1276 = 0.0274
- The overall accuracy rate is 1241/1276 = 0.9725

2.5.5 Area under the curve and receiver operating characteristic

Area under the curve (AUC) and receiver operating characteristic (ROC) curve is used to measure the performance of a classifier in a classification problem at various threshold settings. ROC is a probability curve, whereas AUC tells us about measure or degree of separability. It also tells us the way to distinguish between the classes. If the AUC is higher, the better it predicts.

		Predicted class	
		affinity_card = 1	affinity_card = 0
Actual class	affinity_card = 1	516	25
	affinity_card = 0	10	725

Fig. 3 Example for confusion matrix [7].

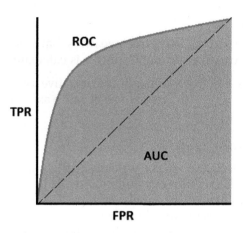

Fig. 4 ROC curve [7].

Fig. 4 shows an example for ROC curve. Here, TPR is the true positive rate and FPR is the false positive rate.

2.6 Classification models for predicting diseases

Several works have been done to build classification models for prediction of diseases. Some of the works are discussed in this section.

Reddy and Khare [4] used a fuzzy rule-based classification algorithm for classification of heart diseases. In their work, they used the publicly available UCI heart disease repository for training the model. They used a principal component analysis (PCA) algorithm for feature selection. Then the extracted features were used for training the algorithm using the fuzzy rule-based classifier.

Reddy et al. [6] used classification to predict the risk in both synthetic and real datasets.

Gadekallu and Khare [8] used a hybrid rough sets and cuckoo search algorithm for feature extraction. Then they used UCI heart disease datasets and a diabetes dataset for training the fuzzy rule-based classification algorithm.

Reddy and Khare [9] used PCA for feature extraction. They used UCI heart disease datasets for training the model. Classification was performed by hybridizing a fuzzy rule-based algorithm with oppositional-based learning, and firefly and bat algorithms.

Thippa Reddy and Khare [10] used diabetes datasets for training the model. They used PCA for feature extraction. The classification model is hybridized with firefly and bat algorithms with fuzzy rule-based classifier. Reddy and Khare used diabetes datasets for classification. The feature extraction was performed via PCA algorithm. The classification model was built by artificial neural networks.

To improve the spatial resolution of EEG data, Han et al. [11] proposed deep convolutional networks. During simulation of Gaussian noise, the authors observed that the super resolution (SR) altered the signal from low resolution (LR) to high resolution (HR). In addition, it improved signal quality. In the real (colored) noise, it recovered the signal to the level of its target data. Even when the upscaling ratio of SR increased, the signal quality obtained was acceptable.

Esteva et al. [12] presented deep learning techniques for health care, centering their discussion on deep learning in computer vision, natural language processing, reinforcement learning, and generalized methods.

Pradeep and Naveen [13] analyzed EHRs to predict the survival rate in lung cancer. An ensemble of support vector machine, Naive Bayes (NB), and classification trees (C4.5) can be used to evaluate patterns that are risk factors for lung cancer. The North Central Cancer Treatment Group (NCCTG) lung cancer dataset along with new patient data was used for evaluating the performance of support vector machine, NB, and C4.5. The comparison was based on accuracy, AUC, and ROC.

Kumar et al. [14] proposed a new classification algorithm called fuzzy rule-based neural classifier for diagnosing disease and disease severity. The experiments used the standard UCI repository dataset and real health records collected from various hospitals.

In 1999, Pendharkar et al. [15] diagnosed breast cancer using diverse association and classification approaches under the data mining technique. In 2001, Richards et al. [16] proposed the data mining technique of association rule to indicate early mortality by analyzing diabetes patients and their clinical records. In 2005, Delen et al. [17] developed three data mining approaches of artificial neural network, decision tree, and logistic regression for predicting breast cancer using a large dataset with various attributes.

In 2008, Abe et al. [18] used a categorized and integrated data mining technique and suggested a framework called Cyber Integrated Medical Infrastructure (CIMI) for medical diagnosis purposes. In 2011, Yeh et al. [19] performed prediction of hemodialysis patients under hospitalization, In 2012, Lahsasna et al. [20] proposed a fuzzy rule-based system in order to diagnose coronary heart disease (CHD) by the proper decision-making system through data mining. The main objectives of the experiment were accuracy of the decision and transparency, which were made possible by a multiobjective genetic algorithm.

In 2015, Sung et al. [21] brought about two data mining techniques with the combination of existing multilinear regression to predict the stroke severity index (SSI) of patients. In 2016, Arslan et al. [22] suggested various data mining techniques such as support vector machine, stochastic gradient boosting (SGB), and penalized logistic regression (PLR) to predict ischemic stroke. In 2016, Chen et al. [23] predicted the personal health index of patients from a geriatric medical examination (GME) dataset using a new technique called MyPHI. Exarchos et al. [24] developed the EMBalance Decision Support System (DSS) for diagnosing balance disorders using a dataset that comprises detailed patient history, disease-related history, medical test results, and other variables.

In 2002, Abbass [25] predicted breast cancer using an evolutionary artificial neural network (EANN) based on the Pareto differential evolution (PDE) algorithm. Galan et al. [26] predicted the spread of nasopharyngeal cancer by suitable classification using a Bayesian network. In 2004, Bojarczuk et al. [27] implemented a genetic programming algorithm for the classification of medical data with breast cancer, chest pain, dermatology, and pediatric adrenocortical tumor and compared the classification using C4.5 and a decision tree. In 2006, Goncalves et al. [28] suggested a new method for rule extraction and classification using inverted hierarchical neuro-fuzzy BSP.

Diabetes, breast cancer, Parkinson's disease, liver disorders, and bladder cancer were diagnosed automatically by Li et al. [29] using feature extraction by PCA and class cation by SVM in 2011. Also in 2011, Gagliardi [30] proposed an instance-based classifier called prototype exemplar learning classifier (PEL-C) to diagnose diseases such as erythemato-squamous disease, heart disease, and diabetes mellitus. In the same year, Astrom and Koker [31] adopted the parallel neural network model to predict Parkinson's disease.

Hospital-acquired pressure ulcers (HAPUs) are predicted by the effective classification process using a Bayesian network model by Cho et al. [32] in 2013. Later in the same year, Di Noia et al. [33] predicted kidney disease at advanced stages using an artificial neural network. In 2016, Weng et al. [34] investigated the different types of neural network classifiers to diagnose major diseases such as heart disease, diabetes, Parkinson's disease, hepatitis, breast cancer, and liver diverse was diagnosed by Bashir et al. [35] using multilayer classifier.

Many researchers have used machine learning approaches along with soft computing techniques and other metaheuristic algorithms to classify text and other datasets like images, diseases, and so on [36–56].

3 Regression model for disease prediction

Predictive models are predominantly used for predicting heart disease, diabetes, and epidemiological disease. In India, these three types of diseases are common. In general, the predictive models in health care should follow the steps outlined in the sections that follow (Fig. 5).

The first phase involves various predictive analytics techniques that predict events in the future. The monitoring phase visualizes the events happening now through dashboards and score cards. The third phase of analysis denotes the inference about the events that occurred using visualization tools. The reporting step deals with query reporting and search tools about the events that happened. The typical predictive model involves the following steps as shown in Fig. 6.

The predictive model can be designed in two phases. The first phase is construction of the predictive model based on historical data and the second phase is validating the model and training it to minimize error. The dataset based on the application considered is taken for the study. Generally, 80% and 20% are considered for training and testing. The first phase deals more with data preprocessing. Information relevant to analysis, dimensionality reduction, and feature extraction are the most important tasks in preprocessing. Various algorithms like support vector machine,

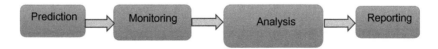

Fig. 5 Steps involved in prediction of healthcare industry.

Fig. 6 Phase-1: Building predictive model.

liner regression, multiple linear regression, k-nearest neighbor, ANN, machine learning, and deep learning are some of the popular algorithms used to build predictive models. Once the predictive model is constructed, the next phase is to validate the model and train it accordingly.

The phase 2 predictive model validates the correctness of the predicted result as shown in Fig. 7. Test data is given to the predictive model and the result is validated using measures including specificity, sensitivity, and ROC curve. The models can be compared with the existing model to prove the efficiency of the model.

3.1 Prediction of heart-related diseases

Coronary artery disease causes many deaths and, as such, early detection of heart disease is essential. Most heart attacks start slowly with mild pain and discomfort, but some attacks are sudden and severe. The death rate associated with heart attacks has increased in developing countries due to lack of understanding of symptoms as well as emotional factors. Treatment success depends on early detection of disease. A prediction model should predict heart disease only based upon patient clinical history factors rather than on test results. Yadav et al. [57] suggested that some of the important risk factors for heart disease are age, sex, family history, smoking, weight, cholesterol, diabetes, and blood pressure. Therefore there is an urgent need for a heart disease prediction system. This predictive model would help patients through patient-oriented software applications or doctors through a decision support system.

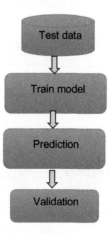

Fig. 7 Phase 2: Predictive model validation.

Ischemic heart disease (IHD) is the most common type of heart disease. Some of the important factors that cause this disease are consumption of fatty food, change in lifestyle, stress, and lack of exercise. A logistic regression model is used to investigate the factors that influence heart disease risk. The dependent attribute is a binary variable that depends on the independent variable, which may be continuous or discrete. Logistic regression finds the probability of prevalence of disease among patients, which can be written as

$$prob(Y = 1) = \frac{e^z}{1 + e^z}$$

where Y is the dependent variable, e is the base of natural logarithms, and Z is given by

$$Z = \beta_0 + \beta_1 X_1 + \beta_2 X_2 + \ldots \beta_p X_p$$

where $\beta_0 \ldots \beta_p$ are coefficients and $X_1 \ldots X_p$ are independent variables for p predictors.

Bhatti et al. [58] implemented a logistic regression model for a study done at Chandka Medical College in Pakistan in 1998. Among 585 observations, there were 101 patients without IHD and 484 patients with IHD. As a result, eight individuals were misclassified in the dataset, which shows that the model fits the IHD data perfectly.

Soleimani and Neshati [59] suggested that the independent variable can be considered for two types of data: clinical test reports and symptoms of the patient. Some of the symptoms of heart attack are pain in the chest and shoulders, drowsiness, weakness, and so on, which the model can use to predict the risk of heart attack. A system like a personal digital assistant (PDA) could be designed for people to estimate the likelihood of attacks based on the symptoms entered in the PDA. Patients could also use the device to obtain the advice of a healthcare professional. The accuracy of the symptoms experienced will obviously affect the performance of the prediction model. Evaluation of the predictive model is required to identify if patients can be correctly classified.

Multiple linear regression is a statistical model that is used to describe the relationship between a dependent variable and more than one independent variables. Analyzing the dependency of variables, fitting the line, and evaluating the accuracy of the model are the various stages of this model [60]. The regression line denotes the probability of heart disease for a given combination of input attributes. The deviation between the input points and regression line is known as residual. This deviation of points can be minimized by least square method.

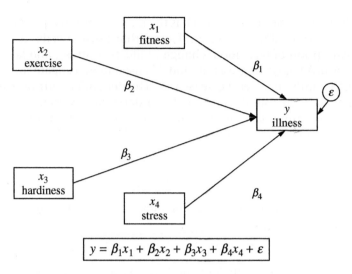

Multiple regression

3.2 Regression models for diabetes mellitus

Diabetes mellitus (DM) is a metabolic disease that depends on level of insulin secretion. Diabetes is the seventh leading cause of death in the United States. Diabetes is a group of diseases characterized by high blood sugar. When a person has diabetes, the body either does not make enough insulin (type 1) or is unable to properly use insulin (type 2). When the body does not have enough insulin or cannot use it properly, sugar builds up in the blood. People with diabetes can develop high blood pressure and high cholesterol and triglycerides (lipids). High blood sugar, particularly when combined with high blood pressure and lipids, can lead to heart disease, stroke, blindness, kidney failure, amputations of the legs and feet, and even early death.

Diabetes is a serious disease that can often be managed through physical activity, diet, and use of insulin and oral medications to lower blood sugar levels. As many as 2 out of 5 Americans are expected to develop type 2 diabetes in their lifetime. Prediabetes is a serious health condition in which a person's blood sugar levels are higher than normal, but not high enough yet to be classified as type 2 diabetes. Without weight loss (for those who need it), healthy eating, and moderate physical activity, many people living with prediabetes will go on to develop type 2 diabetes.

Data mining methods support healthcare researchers to retrieve novel knowledge from large health data. With the development of information technology, data mining and machine learning offer advantages in diabetes research, which leads to improved healthcare distribution, increased support for decision making, and improved disease supervision. Diabetes is a major health problem in India. There is a long history of diabetic registries and databases that systematically collect patient information. This disease has many side effects such as risk of eye disease, risk of kidney failure, and other complications. However, early detection and proper care management can make a big difference.

More people are developing diabetes during youth, and racial and ethnic minorities continue to develop this condition at greater rates. Likewise, the number of older people in our nation is increasing, and older people are more likely to have a chronic disease like diabetes. It's critically important that efforts are intensified to control and manage diabetes and prevent type 2 diabetes. And by addressing diabetes, many other related health problems can be addressed and/or prevented. Therefore, as mentioned, diabetes needs early prevention and diagnosis to save human lives from early death. By considering the seriousness of this disease, various regression models can be used to predict is as early as possible.

In 2015, the International Diabetes Federation (IDF) recorded 415 million diabetes patients. It is predicted that this number will increase to 642 million or 1 in 10 people. Several researchers have used the Pima Indian Diabetes dataset to research diabetes and WEKA as a primary toolkit to implement the model. Patil [61] designed a hybrid approach called the hybrid prediction model (HPM), which uses a k-means clustering algorithm, to validate the selected class label. Its predictive accuracy was found to be 92.38%. Ahmad and Mustaphah [62] made a comparative study based on multilayer perceptron (MLP) with ID2 and J48. The results showed that the prediction accuracy of J48 performed better than the other two algorithms. Marcano et al. [63] improved the accuracy of MLP using artificial metaplasticity and obtained prediction accuracy of 89.93%.

In order to obtain the best accuracy in predicted results, Vijayan and Anjali [64] stated that preprocessing of data before prediction plays a major role. They reviewed different preprocessing techniques for predicting DM such as PCA and discretization. They concluded that the prediction algorithms obtained good accuracy by using preprocessing algorithms, whereas the accuracy of prediction algorithms without preprocessing was decreased. Wei et al. [65] analyzed the risk factors related to DM using Apriori and FP-growth algorithms. Guo [66] validated and tested the predicted model using sensitivity, specificity, and ROC curve. Sowjanya [67] used the DM prediction algorithm in such a way that the model can be useful for everyone. An Android application-based model has been designed to bring about awareness of DM. The application used a decision tree classifier to predict diabetic level and to provide helpful suggestions for diabetics.

Gang et al. [68] developed a mobile-based product that calculates risk of developing DM. Chandrakar [69] designed a diabetes screening tool called Indian Weighted Diabetes Risk Score (IWDRS) to address the issue of late diagnosis. Han and Luo [70] improved the K-means algorithm with pair-wise and size-constrained K-means (PSCKmeans) to find people at high risk of acquiring DM.

Probabilistic prediction models are applied in many areas of biological sciences. This type of model can be used when we need the data to be predicted under various categories. Usually in logistic regression, the target variable is binary and indicates the absence or presence of the disease. The main objective of a logistic regression algorithm is to find the best fit that describes the relationship between the target and predictor variables. To predict the probability of a patient who belong to class "1" for having diabetes is

$$P(y=1|x) = h_\theta(x) = 1 \Big/ 1 + \exp\left(-\theta^T x\right)$$

To predict the probability of a patient who belongs to class "0" for not having diabetes is

$$P(y=0|x) = 1 - h_\theta(x)$$

Some of the other factors that affect the prediction factor are age, sex, marital status, education level, smoking, alcohol consumption, physical activity, and stress (Fig. 8). Diabetes is associated with severe diseases such as heart attack, stroke, retinopathy, and kidney disease. Therefore early prediction can help save lives and improve patient health.

3.3 Prediction models in epidemiology

An epidemic is a disease that spreads in particular location for a specific period of time in a given population. Modeling in epidemics describes the transmission of disease among individuals

Fig. 8 Factors affecting diabetes.

[71]. Prediction research in medicine and epidemiology uses a Bayesian framework and traditional statistical regression technique. Recently, data mining and machine learning are being used more to predict the intensity of disease. With high mortality and uncertain spread of disease, policymakers find it difficult to make decisions on preparedness for the outbreak of disease. Mathematical modeling of an epidemic has a vital role in understanding the various dimensions of underlying epidemiological patterns. There are many mathematical models used to predict the outbreak of disease.

3.3.1 Types of epidemic models

Epidemic models are used to study the mechanism by which diseases spread, predict the pattern of outbreak, and evaluate strategies to control an epidemic [72]. There are two types of epidemic models:
(1) Stochastic models
(2) Deterministic models

3.3.2 Stochastic models

Stochastic models [73] consider the movement of individuals between different classes and not average rate between classes. Addy et al. stated that the stochastic model was developed by Ball in 1986 who discussed that the distribution of infectious period is allowed to have any kind of distribution that can be described by its Laplace transform. Stochastic models are built around random graphs. The stochastic process is the study of how a random variable evolves over time [74]. A variable that is not known before a certain time t is called a random variable. If the state of the random variable is known before a finite time it is called a discrete stochastic process. If the state of the random variable is known at any point of time it is called a continuous stochastic process. The Markov chain process is the best example of a stochastic model where the probability distribution of time $t + 1$ depends on the state at time t and does not depend on the states before time t.

3.3.3 Compartmental model

Brauer [75] discussed various compartmental models built around differential equations. Various terminologies used during compartmental modeling are given in the sections that follow.

Infection rate

The number of people infected within a specific time period

$$\text{Rate of infection} = \frac{\text{Number of infections}}{\text{Number of people in population who are at high risk}}$$

Recovery rate

The number of people who recover from the infection is known as recovery rate

$$\text{Rate of recovery} = \frac{\text{Number of recovered}}{\text{Number of people who are infected}}$$

Latent rate

The number of people who are infected and become infectious is called the latent rate. Even though a person has the pathogen, it takes some time to spread the disease to other persons.

Reproduction number

Reproduction number is used to measure the spread of disease. It is the average number of secondary infection produced by infected people in the population.

$$\text{Reproduction number} = \frac{\text{Infection rate}}{\text{Recovery rate}}$$

Susceptible

People in the population who have high probability of getting infected.

Infected

Infected people who are not prepared to spread the disease in the population.

Exposed

Infected but recovered people who can spread the disease in the population.

In 1927, Kermack and Mckendrick [76] proposed a mathematical model for epidemic disease called susceptible-infectious-recover (SIR) that introduced variables such as rate of infection and recovery. This model served as a starting point to various studies of epidemic models.

The basic SIR model was designed (Fig. 9) to explain the behavior of plague and cholera epidemics. The initial total population is considered in the susceptible compartment, people who got infected either through contact or airborne is taken in the infectious compartment, and individuals recovered from the disease who gained further immunity from it are taken in the recover compartment.

S = fraction of *susceptible* people in a population.
I = fraction of *infected* people in a population.
R = fraction of *recovered* people in a population.

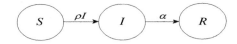

Fig. 9 SIR compartmental model.

The equivalent differential equation for the SIR compartmental model is

$$\frac{dS}{dt} = -\rho SI$$

$$\frac{dI}{dt} = \rho SI - \alpha I$$

$$\frac{dR}{dt} = \alpha I$$

where,
ρ = infection rate
α = removal rate

Li and Muldowney [77] proposed the susceptible-exposed-infected-recover model (SEIR), which includes more dynamics in the system as depicted in Fig. 10. People who are infected from the susceptible population will not become infectious immediately. The latent time between the infected and infectious stage is shown as exposed compartment. In a closed population with no birth or deaths, the equation of SEIR model becomes:

$$\frac{dS}{dt} = -\frac{\beta SI}{N}$$

$$\frac{dE}{dt} = \frac{\beta SI}{N} - \sigma E$$

$$\frac{dI}{dt} = \sigma E - \gamma I$$

$$\frac{dR}{dt} = \gamma I$$

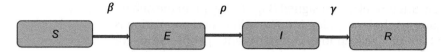

Fig. 10 SEIR model.

where N = S + E + I + R is the total population.

The number of people getting infected per time period is called the infection rate, and the number of people recovered from an infected state is called the removal rate. The important concepts in predicting outbreaks are basic reproduction number denoted by R_0 and generation time (spread of disease from primary case to secondary case). This would also determine the spread of disease. Reproduction number is a quantity defined as expected number of contacts made by an infected person to a susceptible person. The basic reproduction number depends on several factors such as the rate of contacts, probability of infection being transmitted, and the duration of infection.

If $R_0 > 1$, the disease can enter into the entire susceptible population and hence the number of cases will increase. If $R_0 < 1$, the disease will fail to spread in the population, and if $R_0 = 1$, then the disease becomes endemic. Hence reproduction number depicts the pattern of spread of disease. This proves that the main objective of epidemiological modeling is to reduce the reproduction number to less than 1. Prediction of disease spread using compartmental models can help health administrators to make policy decisions such as vaccination strategies to control the spread of disease.

4 Challenges in predicting disease

Algorithms developed for predicting future disease should address the following issues.

(1) Disease models should fit for all types of datasets. Models should not be data specific.
(2) Algorithms that detect disease earlier must be designed to reduce the risk among the population. Early detection of disease can save lives.
(3) The huge amount of medical data should be preprocessed due to the advancement [78] in medical devices, interconnectivity of medical things and patients (Internet of Medical Things), among others. [78]

(4) Risk assessment should consider all the factors that affect the disease.
(5) Decision-making policies can be effectively designed to bring awareness of the disease to people.
(6) Developing applications on health management using mobile devices is more convenient and effective for today's generation.

5 Conclusion

This chapter discussed various classification and prediction techniques and methods. It explained classification methods and predictive models. Classification techniques were explained to provide insight into the different classification models and how to construct and validate them. Predictive models were built for various diseases like heart disease, diabetes, and epidemiological disease.

References

[1] U. Fayyad, G. Piatetsky-Shapiro, P. Smyth, From data mining to knowledge discovery in databases, AI Mag. 17 (3) (1996) 37.
[2] H.C. Koh, G. Tan, Data mining applications in healthcare, J. Healthc. Inf. Manag. 19 (2) (2011) 65.
[3] H. Leopord, W.K. Cheruiyot, S. Kimani, A survey and analysis on classification and regression data mining techniques for diseases outbreak prediction in datasets, Int. J. Eng. Sci. 5 (9) (2016) 1–11.
[4] G.T. Reddy, N. Khare, Heart disease classification system using optimised fuzzy rule based algorithm, Int. J. Biomed. Eng. Technol. 27 (3) (2018) 183–202.
[5] S. Tulsyan, B. Bahl, S. Kaya, G.T. Reddy, Earthquake analysis: visualizing seismic data with python, in: Computational Vision and Bio Inspired Computing, Springer, Cham, 2018, pp. 1041–1048.
[6] G.T. Reddy, A. Srivatsava, K. Lakshmanna, R. Kaluri, S. Karnam, G. Nagaraja, Risk prediction to examine health status with real and synthetic datasets, Biomed. Pharm. J. 10 (4) (2017) 1897–1903.
[7] J. Han, J. Pei, M. Kamber, Data Mining: Concepts and Techniques, Elsevier, 2011.
[8] T.R. Gadekallu, N. Khare, Cuckoo search optimized reduction and fuzzy logic classifier for heart disease and diabetes prediction, Int. J. Fuzzy Syst. Appl. 6 (2) (2017) 25–42.
[9] G.T. Reddy, N. Khare, An efficient system for heart disease prediction using hybrid OFBAT with rule-based fuzzy logic model, J. Circuits Syst. Comput. 26 (04) (2017) 1750061.
[10] G. Thippa Reddy, N. Khare, FFBAT-optimized rule based fuzzy logic classifier for diabetes, Int. J. Eng. Res. Afr. 24 (2016) 137–152 Trans Tech Publications.
[11] S. Han, M. Kwon, S. Lee, S.C. Jun, Feasibility study of EEG super-resolution using deep convolutional networks, in: 2018 IEEE International Conference on Systems, Man, and Cybernetics (SMC), IEEE, 2018, October pp. 1033–1038.

[12] A. Esteva, A. Robicquet, B. Ramsundar, V. Kuleshov, M. DePristo, K. Chou, J. Dean, A guide to deep learning in healthcare, Nat. Med. 25 (1) (2019) 24.

[13] K.R. Pradeep, N.C. Naveen, Lung cancer survivability prediction based on performance using classification techniques of support vector machines, C4. 5 and naive Bayes algorithms for healthcare analytics, Procedia Comput. Sci. 132 (2018) 412–420.

[14] P.M. Kumar, S. Lokesh, R. Varatharajan, G.C. Babu, P. Parthasarathy, Cloud and IoT based disease prediction and diagnosis system for healthcare using fuzzy neural classifier, Futur. Gener. Comput. Syst. 86 (2018) 527–534.

[15] P.C. Pendharkar, J.A. Rodger, G.J. Yaverbaum, N. Herman, M. Benner, Association, statistical, mathematical and neural approaches for mining breast cancer patterns, Expert Syst. Appl. 17 (3) (October 1999) 223–232.

[16] G. Richards, V.J. Rayward-Smith, P.H. Sönksen, S. Carey, C. Weng, Data mining for indicators of early mortality in a database of clinical records, Artif. Intell. Med. 22 (3) (June 2001) 215–231.

[17] D. Delen, G. Walker, A. Kadam, Predicting breast cancer survivability: a comparison of three data mining methods, Artif. Intell. Med. 34 (2) (June 2005) 113–127.

[18] A. Abe, N. Hagita, M. Furutani, Y. Furutani, R. Matsuoka, Categorized and integrated data mining of medical data from the viewpoint of chance discovery, Knowl.-Based Intell. Inform. Eng. Syst. 6278 (2008) 307–314.

[19] J.-Y. Yeh, T.-H. Wu, C.-W. Tsao, Using data mining techniques to predict hospitalization of hemodialysis patients, Decis. Support. Syst. 50 (2) (January 2011) 439–448.

[20] A. Lahsasna, R.N. Ainon, R. Zainuddin, A. Bulgiba, Design of a fuzzy-based decision support system for coronary heart disease diagnosis, J. Med. Syst. 36 (5) (October 2012) 3293–3306.

[21] S.-F. Sung, C.-Y. Hsieh, Y.-H.K. Yang, H.-J. Lin, C.-H. Chen, Y.-W. Chen, Y.-H. Hu, Developing a stroke severity index based on administrative data was feasible using data mining techniques, J. Clin. Epidemiol. 68 (11) (November 2015) 1292–1300.

[22] A.K. Arslan, C. Colak, M.E. Sarihan, Different medical data mining approaches based prediction of ischemic stroke, Comput. Methods Prog. Biomed. 130 (July 2016) 87–92.

[23] L. Chen, L. Xue, Y. Yang, H. Kurniawati, Q.Z. Sheng, H.-Y. Hu, N. Huang, Personal health indexing based on medical examinations: a data mining approach, Decis. Support. Syst. 81 (January 2016) 54–65.

[24] T.P. Exarchos, G. Rigas, A. Bibas, D. Kikidis, C. Nikitas, F.L. Wuyts, B. Ihtijarevic, L. Maes, M. Cenciarini, C. Maurer, N. Macdonald, D.-E. Bamiou, L. Luxon, M. Prasinos, G. Spanoudakis, D.D. Koutsouris, D.I. Fotiadis, Mining balance disorders' data for the development of diagnostic decision support systems, Comput. Biol. Med. 77 (October 2016) 240–248.

[25] H.A. Abbass, An evolutionary artificial neural networks approach for breast cancer diagnosis, Artif. Intell. Med. 25 (3) (July 2002) 265–281.

[26] S.F. Galan, F. Aguado, F.J. Dı́ez, J. Mira, NasoNet, modeling the spread of nasopharyngeal cancer with networks of probabilistic events in discrete time, Artif. Intell. Med. 25 (3) (2002) 247–264.

[27] C.C. Bojarczuk, H.S. Lopes, A.A. Freitas, E.L. Michalkiewicz, A constrained-syntax genetic programming system for discovering classification rules: application to medical data sets, Artif. Intell. Med. 30 (1) (January 2004) 27–48.

[28] L.B. Goncalves, M.M.B.R. Vellasco, M.A.C. Pacheco, F.J. de Souza, Inverted hierarchical neuro-fuzzy BSP system: a novel neuro-fuzzy model for pattern classification and rule extraction in databases, IEEE Trans. Syst. Man Cybern. Part C Appl. Rev. 36 (2) (March 2006) 236–248.

[29] D.-C. Li, C.-W. Liu, S.C. Hu, A fuzzy-based data transformation for feature extraction to increase classification performance with small medical data sets, Artif. Intell. Med. 52 (1) (May 2011) 45–52.
[30] F. Gagliardi, Instance-based classifiers applied to medical databases: diagnosis and knowledge extraction, Artif. Intell. Med. 52 (3) (2011) 123–139.
[31] F. Astrom, R. Koker, A parallel neural network approach to prediction of Parkinson's disease, Expert Syst. Appl. 38 (10) (September 2011) 12470–12474.
[32] I. Cho, I. Park, E. Kim, E. Lee, D.W. Bates, Using EHR data to predict hospital-acquired pressure ulcers: a prospective study of a bayesian network model, Int. J. Med. Inform. 82 (2013) 1059–1067.
[33] T. Di Noia, V.C. Ostuni, F. Pesce, G. Binetti, D. Naso, F.P. Schena, E. Di Sciascio, An end stage kidney disease predictor based on an artificial neural networks ensemble, Expert Syst. Appl. 40 (11) (September 2013) 4438–4445.
[34] C.-H. Weng, T.C.-K. Huang, R.-P. Han, Disease prediction with different types of neural network classifiers, Telematics Inform. 33 (2) (May 2016) 277–292.
[35] S. Bashir, U. Qamar, F.H. Khan, L. Naseem, HMV: a medical decision support framework using multi-layer classifiers for disease prediction, J. Computat. Sci. 13 (2016) 10–25.
[36] T.R. Sooraj, R.K. Mohanty, B.K. Tripathy, Fuzzy soft set theory and its application in group decision making, in: Advanced Computing and Communication Technologies, Springer, Singapore, 2016pp. 171–178.
[37] B.K. Tripathy, T.R. Sooraj, R.K. Mohanty, A new approach to fuzzy soft set theory and its application in decision making, in: Computational Intelligence in Data Mining, vol. 2, Springer, New Delhi, 2016pp. 305–313.
[38] B.K. Tripathy, A. Tripathy, K.G. Rajulu, Possibilistic rough fuzzy C-means algorithm in data clustering and image segmentation, in: 2014 IEEE International Conference On Computational Intelligence and Computing Research, IEEE, 2014pp. 1–6.
[39] B.K. Tripathy, A. Tripathy, K. Govindarajulu, R. Bhargav, On kernel based rough intuitionistic fuzzy C-means algorithm and a comparative analysis, in: Advanced Computing, Networking and Informatics, vol. 1, Springer, Cham, 2014pp. 349–359.
[40] S. Purushotham, B. Tripathy, A comparative study of RIFCM with other related algorithms from their suitability in analysis of satellite images using other supporting techniques, Kybernetes 43 (1) (2014) 53–81.
[41] B.K. Tripathy, A. Ghosh, SDR: an algorithm for clustering categorical data using rough set theory, in: 2011 IEEE Recent Advances in Intelligent Computational Systems, IEEE, 2011pp. 867–872.
[42] B.K. Tripathy, T.R. Sooraj, R.K. Mohanty, A new approach to interval-valued fuzzy soft sets and its application in decision-making, in: Advances in Computational Intelligence, Springer, Singapore, 2017pp. 3–10.
[43] B.K. Tripathy, A. Mitra, An algorithm to achieve k-anonymity and l-diversity anonymisation in social networks, in: 2012 Fourth International Conference on Computational Aspects of Social Networks (CASoN), IEEE, 2012pp. 126–131.
[44] A. Gupta, B.K. Tripathy, A generic hybrid recommender system based on neural networks, in: 2014 IEEE International Advance Computing Conference (IACC), IEEE, 2014pp. 1248–1252.
[45] S. Purushotham, B.K. Tripathy, Evaluation of classifier models using stratified tenfold cross validation techniques, in: International Conference on Computing and Communication Systems, Springer, Berlin, Heidelberg, 2011pp. 680–690.
[46] J. Anuradha, V. Ramachandran, K.V. Arulalan, B.K. Tripathy, Diagnosis of ADHD using SVM algorithm, in: Proceedings of the Third Annual ACM Bangalore Conference, ACM, 2010, Januaryp. 29.

[47] B.K. Tripathy, Rough sets on intuitionistic fuzzy approximation spaces, in: Proceedings of 3rd International IEEE Conference on Intelligent Systems (IS06), London, 2006, pp. 776–779.

[48] B.K. Tripathy, H.K. Tripathy, Covering based rough equivalence of sets and comparison of knowledge, in: 2009 International Association of Computer Science and Information Technology-Spring Conference, IEEE, 2009pp. 303–307.

[49] B.K. Tripathy, D. Mittal, Hadoop based uncertain possibilistic kernelized c-means algorithms for image segmentation and a comparative analysis, Appl. Soft Comput. 46 (2016) 886–923.

[50] B.K. Tripathy, R.K. Mohanty, T.R. Sooraj, On intuitionistic fuzzy soft set and its application in group decision making, in: 2016 International Conference on Emerging Trends in Engineering, Technology and Science (ICETETS), IEEE, 2016pp. 1–5.

[51] T. Gadekallu, A. Soni, D. Sarkar, L. Kuruva, Application of sentiment analysis in movie reviews, in: Sentiment Analysis and Knowledge Discovery in Contemporary Business, IGI Global, 2019pp. 77–90.

[52] G.T. Reddy, M.P.K. Reddy, K. Lakshmanna, D.S. Rajput, R. Kaluri, G. Srivastava, Hybrid genetic algorithm and a fuzzy logic classifier for heart disease diagnosis, Evol. Intel. (2019) 1–12.

[53] B.K. Tripathy, On approximation of classifications, rough equalities and rough equivalences, in: Rough Set Theory: A True Landmark in Data Analysis, Springer, Berlin, Heidelberg, 2009 pp. 85–133.

[54] B.K. Tripathy, A. Mitra, J. Ojha, On rough equalities and rough equivalences of sets, in: International Conference on Rough Sets and Current Trends in Computing, Springer, Berlin, Heidelberg, 2008 pp. 92–102.

[55] S. Bhattacharya, R. Kaluri, S. Singh, M. Alazab, U. Tariq, A novel PCA-firefly based XGBoost classification model for intrusion detection in networks using GPU, Electronics 9 (2) (2020) 219.

[56] T.R. Gadekallu, N. Khare, S. Bhattacharya, S. Singh, P.K. Reddy Maddikunta, I.H. Ra, M. Alazab, Early detection of diabetic retinopathy using PCA-firefly based deep learning model, Electronics 9 (2) (2020) 274.

[57] P.K. Yadav, K.L. Jaiswal, S.B. Patel, D.P. Shukla, Intelligent heart disease prediction model using classification algorithms, Int. J. Comput. Sci. Mobile Comput. 3 (8) (2013) 102–107.

[58] I.P. Bhatti, H.D. Lohano, Z.A. Pirzado, I.A. Jafri, A logistic regression analysis of the ischemic heart disease risk, J. Appl. Sci. 6 (4) (2006) 785–788.

[59] P. Soleimani, A. Neshati, Applying the regression technique for prediction of the acute heart attack, Int. J. Med. Health Biomed. Bio. Eng. 9 (11) (2015) 767–771.

[60] P. Sivakumar, V. Sarvani, P. Prudhvi Raj, K. Suma, D. Nandu, Prediction of heart disease using multiple regression analysis and support vector machine, J. Adv. Res. Dyn. Control Syst. 9 (2017).

[61] B.M. Patil, R.C. Joshi, D. Toshniwal, Hybrid prediction model for type-2 diabetic patients, Expert Syst. Appl. 37 (12) (2010) 8102–8108.

[62] A. Ahmad, A. Mustapha, E.D. Zahadi, N. Masah, N.Y. Yahaya, Comparison between neural networks against decision tree in improving prediction accuracy for diabetes mellitus, in: International Conference on Digital Information Processing and Communications, Springer, Berlin, Heidelberg, 2011, July pp. 537–545.

[63] A. Marcano-Cedeño, J. Torres, D. Andina, A prediction model to diabetes using artificial metaplasticity, in: International Work-Conference on the Interplay Between Natural and Artificial Computation, Springer, Berlin, Heidelberg, 2011 pp. 418–425.

[64] V.V. Vijayan, C. Anjali, Decision support systems for predicting diabetes mellitus—a review, in: 2015 Global Conference on Communication Technologies (GCCT), IEEE, 2015 pp. 98–103.

[65] Z. Wei, Y.E. Guangjian, N. Wang, Analysis for risk factors of type 2 diabetes mellitus based on FP-growth algorithm, China Med. Equip. 13 (5) (2016) 45–47.

[66] Y.R. Guo, Y.Q. Li, G.S. Wang, X.T. Liu, Application of artificial neural network to predict individual risk of type 2 diabetes mellitus, J. Zhejiang Univ. (Med. Sci. Ed.) 49 (2) (2014) 180–183.

[67] K. Sowjanya, A. Singhal, C. Choudhary, MobDBTest: a machine learning based system for predicting diabetes risk using mobile devices, in: 2015 IEEE International Advance Computing Conference (IACC), IEEE, 2015, Junepp. 397–402.

[68] S. Gang, L. Shanshan, Y. Ding, Design and implementation of diabetes risk assessment model based on mobile things, in: 2015 7th International Conference on Information Technology in Medicine and Education (ITME), IEEE, 2015, Novemberpp. 425–428.

[69] O. Chandrakar, J.R. Saini, Development of Indian weighted diabetic risk score (IWDRS) using machine learning techniques for type-2 diabetes, in: Proceedings of the 9th Annual ACM India Conference, ACM, 2016 pp. 125–128.

[70] L. Han, S. Luo, H. Wang, L. Pan, X. Ma, T. Zhang, An intelligible risk stratification model based on pairwise and size constrained Kmeans, IEEE J. Biomed. Health Inform. 21 (5) (2016) 1288–1296.

[71] D. Mollison, M. Denis (Eds.), Epidemic Models: Their Structure and Relation to Data, In: vol. 5, Cambridge University Press, 1995.

[72] M.J. Keeling, L. Danon, Mathematical modelling of infectious diseases, Br. Med. Bull. 92 (1) (2009) 33–42.

[73] R.W. Mbogo, L.S. Luboobi, J.W. Odhiambo, A stochastic model for malaria transmission dynamics, J. Appl. Math. (2018) 2018.

[74] C.L. Addy, I.M. Longini Jr., M. Haber, A generalized stochastic model for the analysis of infectious disease final size data, Biometrics (1991) 961–974.

[75] F. Brauer, Compartmental models in epidemiology, in: Mathematical Epidemiology, Springer, Berlin, Heidelberg, 2008 pp. 19–79.

[76] W.O. Kermack, A.G. McKendrick, A contribution of the mathematical theory of epidemics, Proc. R. Soc. 115 (1927) 700–772.

[77] M.Y. Li, J.S. Muldowney, Global stability for the SEIR model in epidemiology, Math. Biosci. 125 (2) (1995) 155–164.

[78] G.T. Reddy, M.P.K. Reddy, K. Lakshmanna, R. Kaluri, D.S. Rajput, G. Srivastava, T. Baker, Analysis of dimensionality reduction techniques on big data, IEEE Access 8 (2020) 54776–54788.

Further reading

D. Singh Rajput, A.P. Singh, Y. Agarwal, P.K. Reddy, G. Thippa Reddy, Feature selection analysis for multimedia event detection, in: Materials Science and Engineering Conference Series (vol. 263, No. 4, p. 042004), 2017, November.

G.T. Reddy, N. Khare, Hybrid firefly-bat optimized fuzzy artificial neural network based classifier for diabetes diagnosis, Int. J. Intell. Eng. Syst. 10 (4) (2017) 18–27.

12

Clavicle bone segmentation from CT images using U-Net-based deep learning algorithm

Parita Sanghani[a], Francis Wong[b], and Hongliang Ren[a]
[a]Department of Biomedical Engineering, National University of Singapore, Singapore, Singapore. [b]Department of Orthopedic Surgery, Sengkang General Hospital Singapore, Singapore, Singapore

1 Introduction

Medical images are usually from a variety of sources, including computed tomography (CT) for imaging bone structures in orthopedic procedures. The images obtained from these medical imaging modalities are to envision the anatomical compositions of the human body, such as bones, blood vessels, and the different organs in the body. The most common medical image segmentation tasks involve delineating different anatomical structures (such as different types of bones) and unhealthy tissues (such as brain or breast lesions) [1].

Manually annotating anatomical structures on medical images is prone to errors and a time-consuming task involving interrater variability. Annotation of medical image segmentation in a semiautomatic or fully automatic way is of great value in the community. Hence, studies have tried to obtain accurate and dependable solutions to boost workflow efficiency and support, making better decisions by automating the segmentation process [2].

Since the introduction of convolutional neural networks (CNNs) [3,4], they popularly emerge in tasks of medical image segmentation. The recent advances in semantic segmentation have enabled various organ segmentation, such as cardiac [5] and lung detection [6], with high accuracy using CNNs. CNNs have become the gold standard due to factors such as filter sharing, high representation power, and fast inference. The most commonly used architectures are fully convolutional networks [7] and the U-Net [8].

Segmentation of clavicle bones is crucial, especially for early diagnosis [9,10]. Segmentation of the clavicle bone is essential for clinicians to generate personalized implants for the patient whose clavicle bone is injured due to accident or collision. It can also enable clinicians to perform a shape analysis of the bone for the development of bone shape models and generating personalized clavicle. Most studies have carried out clavicle segmentation using chest radiographs [10–14]. This study aims to perform computer-assisted clavicle bone delineation from CT images using deep learning, where the algorithm takes in the CT image as input and provides segmented clavicle bone as output.

2 Materials and methods

2.1 Data

We performed our study of automated clavicle segmentation on lung CT data of 156 subjects. Each CT image slice has 256 × 256 pixels along with the sagittal and coronal directions. However, it varied in different patients along the axial direction. The clavicle bone was manually segmented by experts voxel by voxel, which was ground truth for this study.

In our study, we employed the two-dimensional U-Net architecture to perform automated clavicle segmentation. Although other deep learning architectures can perform the segmentation of the clavicle bones, we chose to use U-Net as it is a common choice and has performed better than several other networks. It is also a very network.

The training set consisted of randomly chosen 110 patient images (7252 2-D slices along axial direction), while the test set consisted of the remaining 46 patient images (3027 2-D slices along axial direction).

Fig. 1 shows examples of a few slices from our dataset with the clavicle bone segmented manually and marked in red.

2.2 Network architecture

Our U-Net network consists of a down-sampling path, including encoding block/contracting parts, and an up-sampling path, including decoding block/expansive parts. The contracting part of the network consisted of four convolutional blocks. Every block had two convolutional layers with a filter size of 3 × 3, a stride of 1, and rectifier linear activation. The down-sampling path involves

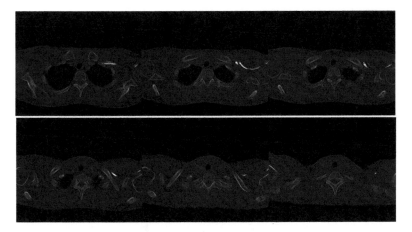

Fig. 1 Examples of a few slices from our training dataset with the clavicle bone segmented manually and annotated in *red (gray in the print version)*.

max pooling with a size of 2 × 2, so the size of feature maps decreased from 256 × 256 (input size) to 32 × 32. Every step in the expansive path consisted of an up-sampling layer with a size of 2 × 2 to double the size of the feature map, tailgated by concatenation with the correspondingly cropped feature map from the contracting path, two 3 × 3 convolution layers with a stride of 1, and a rectified linear activation. Unlike the initial U-Net construction [6], we used zero/same padding to ensure that the output dimension was the same as the input dimension. The final layer, a 1 × 1 convolution, was to map the feature vector to the output number of two classes (clavicle and nonclavicle) with a sigmoid activation function.

Fig. 2 represents the construction of our U-Net Model 1 in our study (referred to as *Model 1* in this study).

We performed another experiment, where we doubled the amount of output filters in every individually convolutional layer in the expansive path of the architecture (illustrated in Fig. 3) (referred to as Model 2 in this study).

No data augmentation was needed in our experiment, as we had a large number of 2-D input slices per patient. A batch size of 10 was to train the network with 100 epochs.

Our network implementation employs Keras with TensorFlow backend and NVIDIA Quadro M4000 8GB 4xDP GPU card. Due to the computational and memory limitations of the GPU card, we chose to implement a 2-D U-Net model. Additionally, it also results in faster inference.

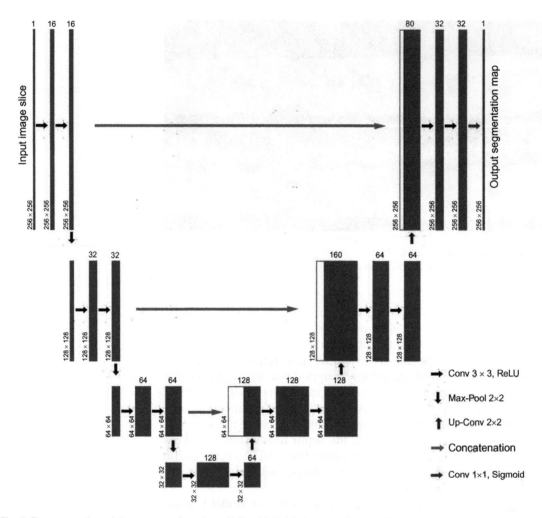

Fig. 2 Representation of the construction of our U-Net Model 1.

2.3 Training and optimization

In our CT slices the clavicle extends only a tiny region of the scan slice. Hence, while trying to minimize the loss function, the neural network is likely to get confined in a local minimum of the loss function. This confinement can cause the network to be strongly biased toward the background and miss out or partially detect the clavicle pixels. To overcome this, our study adopted the dice loss function, which is the following:

$$\text{Dice Loss} = 1 - \text{Dice Similarity Coefficient}$$

$$\text{Dice Similarity Coefficient} = \frac{2*TP}{FP + 2*TP + FN}$$

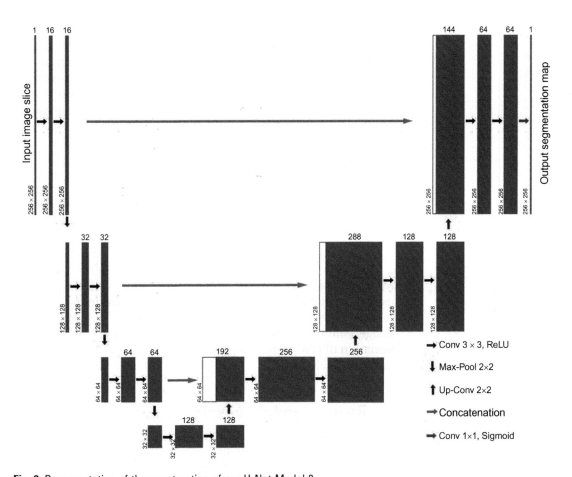

Fig. 3 Representation of the construction of our U-Net Model 2.

where TP, FP, and FN indicate the true positive, false-positive, and false-negative measurements, respectively. The value of DSC ranges between 0 and 1.

Smoothed dice metric, along with root mean squared error (RMSE), was used to judge the performance of the model. Smoothed dice function is a slightly modified version of the primary dice similarity coefficient (DSC) function, due to the addition of a smoothness parameter. The advantage of using a smoothed dice metric is that a regular DSC metric would require a threshold of the output before computation of the loss, whereas smoothed dice has no such requirement. The smoothed dice value is

$$\text{Smoothed Dice} = \frac{2*TP + \text{smooth}}{2*TP + FP + FN + \text{smooth}}$$

where the smoothness parameter was set to 1.0.

Methods like stochastic gradient-based optimization are to train deep neural networks to minimize the cost function. In our study, we used the adaptive moment optimizer/estimator (Adam) [12] with the lower-order moments of gradients to evaluate the parameters. The learning rate parameter of our Adam optimizer was set as 0.00001.

2.4 Computation of accuracy

After obtaining the predicted segmentation of the CT volumes in the test set, we computed the accuracy of both the models using the dice similarity coefficient (DSC). DSC was computed as

$$\mathrm{DSC} = \frac{2|V_1 \cap V_2|}{|V_1 + V_2|}$$

where $V1$ was the volume of the predicted clavicle segmented using the model and $V2$ was the volume of the manually segmented clavicle. The DSC value was computed for every patient image volume for both models. The mean value and standard deviation of DSC were computed of the 46 test patient image volumes for both models.

3 Results and discussion

We trained our model using 110 patient image volumes and their corresponding manually contoured ground truth annotations provided by a clinical expert. This training dataset was reflecting the clinical variability found in clinical contexts. We tested our model on randomly chosen 46 patient image volumes (3027 2-D slices).

Table 1 shows the minimum, maximum, and mean DSC computed using Model 1 and Model 2 from the 46 patient volumes in

Table 1 The minimum, maximum, and the mean DSC computed using Model 1 and Model 2 from the 46 patient volumes in the test dataset.

	Minimum DSC	Maximum DSC	Mean DSC
Model 1	0.62	0.87	0.76
Model 2	0.75	0.95	0.92

the test dataset. For Model 1 the minimum DSC computed was 0.62, the highest DSC value was 0.87, and the mean DSC value was 0.76. For Model 2 the minimum DSC computed was 0.75, the highest DSC value was 0.95, and the mean DSC value was found to be 0.92.

Fig. 4 illustrates the distribution of test set patient images concerning the DSC achieved through the segmentation using Model 1 and Model 2.

Along with the quantitative assessment, we conducted qualitative assessments as well. The images were visualized by overlaying the manual and predicted segmentations as masks using ITK-Snap. We assessed the segmentation of the clavicle bone obtained by both models slice by slice by adjusting the opaqueness. Although Model 1 correctly identified the bone region, it did miss out on the hollow structure inside the bone, even though it was part of the manual segmentation. On the other hand, Model 2 performed much better compared with Model 1 in replicating the ground truth labels and predicting the complete clavicle bone.

Fig. 5 shows examples of a few slices from our dataset with the clavicle bone segmented manually (left) and resulting segmentation using Model 1 and Model 2. In all the slices, Model 1 leads to under segmentation, and Model 2 consistently performed better than Model 1.

Although the quantitative assessment made it clear that the Model 2 outperforms Model 1 based on the DSC values, it was only after the qualitative assessment that we were able to determine precisely where Model 2 was performing better than Model 1.

Our experiment demonstrated that by only doubling the number of convolutional filters and keeping all other parameters the same, the model could more closely replicate the ground truth labels that it is trained.

Our study has some limitations. First, we adopted the 2-D U-Net architecture instead of 3-D, due to computing and memory constraints. Second, several parameters in our models can fine-tune to boost the performance. We did, however, boost our first performance of Model 1 by doubling the number of convolutional filters in Model 2. Third, we performed our study only based on the U-Net model. However, future studies should include other architectures and assess which architecture is best suited for automated clavicle segmentation from CT images. Fourth, our qualitative assessment and visual analysis were by one clinical expert and a biomedical engineer.

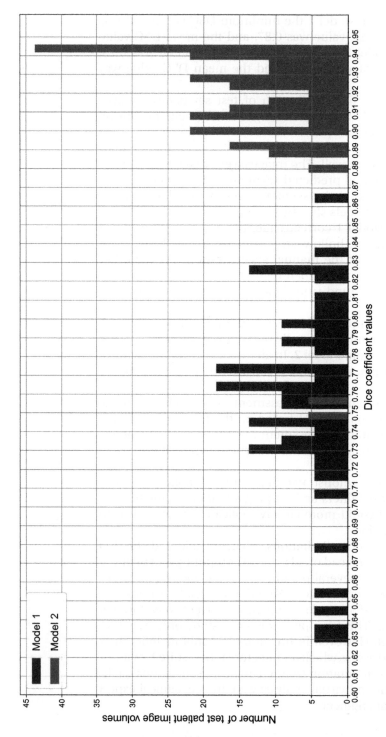

Fig. 4 Illustration of the distribution of DSC across test set patient cases compared between the two models.

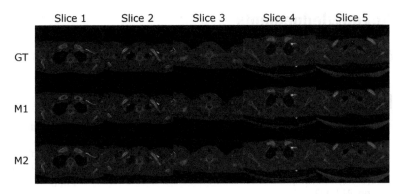

Fig. 5 Example of a few slices from our dataset with the clavicle bone segmented manually (*left*) and resulting segmentation using Model 1 and Model 2. GT refers to the ground truth segmentation as performed by the experts. M1 and M2 refer to the resulting segmentation from Model 1 and Model 2 in our study.

The entire dataset was from the Asian population and the same institution. In the future, models should train on multipopulation and multiinstitutional data to introduce more variance in the dataset.

Despite these limitations the U-Net-based models in our study have demonstrated promising segmentation results with high accuracy. Even with a simple construction of the U-Net, we obtained tremendous accuracy. This precision can be attributed to the limited range of variability of clavicle bones in the dataset, as it all belonged to the Asian population with the images acquired in only one institution. Moreover, our study shows the fact that a simple 2-D U-Net-based architecture can perform clavicle bone segmentation with accuracy.

4 Conclusion

This study demonstrates the application of U-Net-based models for automated clavicle segmentation task. Our models yielded high performance. Our models can be applied in daily practice for clavicle segmentation, which can enhance the process of generating personalized implants and performing shape analysis of the clavicle bone for the development of bone shape models. Our future work involves trying to overcome the limitations of this current work and comparing our architecture with other deep learning networks.

Acknowledgements

This work was supported in part by the NMRC Bedside & Bench under grant R-397-000-245-511, Singapore Academic Research Fund under Grant R-397-000-297-114.

References

[1] B. Kayalibay, G. Jensen, P. van der Smagt, CNN-based segmentation of medical imaging data, arXiv (2017) 1–24 preprint arXiv:1701.03056.
[2] M. Islam, Y. Li, H. Ren, Learning where to look while tracking instruments in robot-assisted surgery, in: International Conference on Medical Image Computing and Computer-Assisted Intervention, October 13, 2019, Springer, 2019, pp. 412–420.
[3] V.E. Balas, S.S. Roy, D. Sharma, P. Samui (Eds.), Handbook of Deep Learning Applications, Springer, 2019.
[4] P. Samui, S.S. Roy, V.E. Balas (Eds.), Handbook of Neural Computation, Academic Press, 2017.
[5] W. Bai, M. Sinclair, G. Tarroni, O. Oktay, M. Rajchl, G. Vaillant, A.M. Lee, N. Aung, E. Lukaschuk, M.M. Sanghvi, F. Zemrak, Human-Level CMR Image Analysis With Deep Fully Convolutional Networks, arXiv (2017).
[6] F. Liao, M. Liang, Z. Li, X. Hu, S. Song, Evaluate the malignancy of pulmonary nodules using the 3-D deep leaky noisy-or network, IEEE Trans. Neural Netw. Learn. Syst. 30 (11) (2019) 3484–3495.
[7] J. Long, E. Shelhamer, T. Darrell, Fully convolutional networks for semantic segmentation, in: Proceedings of the IEEE Conference on Computer Vision and Pattern Recognition, 2015, pp. 3431–3440.
[8] O. Ronneberger, P. Fischer, T. Brox, U-net: convolutional networks for biomedical image segmentation, in: International Conference on Medical Image Computing and Computer-Assisted Intervention, Springer, Cham, 2015, pp. 234–241.
[9] B. Van Ginneken, S. Katsuragawa, B.M. Ter Haar Romeny, K. Doi, M.A. Viergever, Automatic detection of abnormalities in chest radiographs using local texture analysis, IEEE Trans. Med. Imaging 21 (2) (2002) 139–149.
[10] A.A. Novikov, D. Lenis, D. Major, J. Hladůvka, M. Wimmer, K. Bühler, Fully convolutional architectures for multiclass segmentation in chest radiographs, IEEE Trans. Med. Imaging 37 (8) (2018) 1865–1876.
[11] M. Ropars, H. Thomazeau, D. Hutton, Clavicle fractures, Orthop. Traumatol. Surg. Res. 103 (1) (2017) S53–S59.
[12] M. Islam, D.A. Atputharuban, R. Ramesh, H. Ren, Real-time instrument segmentation in robotic surgery using auxiliary supervised deep adversarial learning, IEEE Robot. Autom. Lett. 4 (2) (2019) 2188–2195.
[13] B. Van Ginneken, M.B. Stegmann, M. Loog, Segmentation of anatomical structures in chest radiographs using supervised methods: a comparative study on a public database, Med. Image Anal. 10 (1) (2006) 19–40.
[14] M. Eslami, S. Tabarestani, S. Albarqouni, E. Adeli, N. Navab, M. Adjouadi, Image-to-images translation for multi-task organ segmentation and bone suppression in chest X-ray radiography, IEEE Trans. Med. Imaging 39 (7) (2020) 2553–2565.

13

Accurate classification of heart sounds for disease diagnosis by using spectral analysis and deep learning methods

Pratima Upretee[a] and Mehmet Emin Yüksel[b]

[a]Department of Biomedical Engineering, Graduate School of Natural Sciences, Erciyes University, Kayseri, Turkey. [b]Department of Biomedical Engineering, Faculty of Engineering, Erciyes University, Kayseri, Turkey

1 Introduction

Many deaths around the globe, especially in the underdeveloped and developing countries, are caused due to inadequate diagnostic services. Most people prefer staying at home rather than visiting a doctor in the presence of slight symptoms. This is mainly due to ignorance, but poor economic conditions are another major factor as these patients cannot usually afford expensive healthcare services. Undiagnosed condition or delayed diagnosis of diseases often leads to untimely deaths. Hence, alternatives to expensive and sophisticated health services should be developed to reduce the increasing trend of untimely deaths.

Various physiological signals, such as *electrocardiogram* (*ECG*), *phonocardiogram* (*PCG*), *electromyogram* (*EMG*), and *electroencephalogram* (*EEG*), are used in the diagnosis of different kinds of diseases [1]. The disease diagnosis process involving these types of physiological signals can be automated with the help of biomedical signal processing. In this study, we have selected the *phonocardiogram* (*PCG*) signals, considering the increasing trend of death rate due to cardiovascular diseases (CVDs).

According to the statistical reports made available by World Health Organization (WHO) and several other organizations in the recent years, cardiovascular diseases have continued to be the number one reason for worldwide deaths for more than a decade [2, 3]. Even though there have been many advancements in the field of cardiovascular disease diagnosis, the number of deaths is not showing any sign of decrease. Many sudden cardiac deaths are caused due to delayed diagnosis [4], and therefore an urgent alternative should be developed for the early diagnosis of CVDs.

Normally the first step of the cardiac disease examination is the *cardiac auscultation*, where a physician listens to the heart sound of the patient with the help of a stethoscope. Stethoscopes have been continuously used in diagnostic procedures for more than 200 years, and they are very simple, noninvasive, cheap, and portable devices [5, 6]. Nowadays, digital stethoscopes are available allowing cardiac sounds to be recorded during the examination for further analysis. These recordings are known as *phonocardiograms*, and the process is known as *phonocardiography*.

Cardiac sounds contain useful information related to the pathology of the heart. Hence, cardiac auscultation can be used as an alternative solution to the expensive and complex heart diagnosis systems. It is noninvasive, time efficient, and very simple to record and store heart sounds for further analysis [1, 5]. However, deciphering the information from phonocardiograms is not an easy task. The heart disease diagnosis through auscultation demands higher expertise from the physicians [5–8].

Alongside the morphological features found in time domain, there are many other features hidden in different domains such as frequency domain and time-frequency domain [9, 10]. Manually extracting these hidden features is extremely difficult and time taking [1]. Therefore, there is a need for automatic analysis of heart sound signals to extract all the features regarding the condition of the heart.

Feature extraction is one of the vital components of an automatic heart sound classification system. The accuracy of the classifier is dependent on the quality of the features extracted. There are different methods used for extracting features. The feature extraction methods are selected in accordance to the domain being analyzed.

The feature extraction from time domain is the simplest job as compared with other domains. It includes morphological patterns, positions of components of the heart sound, its amplitude, peak, and other statistical values [9–11].

Nevertheless, time-domain features may not represent all the required features from the heart sound recording. To undermine more quality features, frequency domain analysis with the help of methods, such as discrete Fourier transform, and autoregressive modeling, are opted [8, 12, 13].

Currently, researchers have inclined toward the time-frequency analysis for cardiac sounds since it provides a platform to inspect the features related to both time domain and frequency domain at the same time [14]. Various methods such as Fourier transform, wavelet transform, and short-time Fourier transform (STFT) are employed for this objective [14–19].

Among them, STFT is very popular in automated cardiac auscultation systems [6, 20–24]. However, there is a limitation in resolution of the obtained features due to the width of the window used in this method. A narrower window width means higher resolution in time domain and lower resolution in frequency domain and vice-versa [25].

To solve the shortcomings of STFT, wavelet transform, where time or frequency scaling operation replaces the frequency shifting operation of STFT, is used [26]. Wavelet transform methods are used widely in the cases where the frequency spectrum of signal varies with the time. They are very popular in multiresolution analysis because the width of the temporal window can be altered together with the frequency [20, 22–24, 27]. Various forms of wavelet analysis such as discrete wavelet transform, continuous wavelet transform, and Mel-scaled wavelet transform are currently in use for heart sound analysis [6, 14, 16, 17, 20, 22–24, 27–30].

Similarly, there are other approaches to define the characteristics of heart sound like Shannon energy envelope and linear prediction coefficient. The perceptual features such as Mel-frequency cepstral coefficient are sometimes combined with other types of features to detect and classify the malfunctions in the heart valves through heart sounds [15, 16, 31–34].

The final stage of an automatic cardiac auscultation system is the stage of classification. Here, the extracted features will be used as the input to an appropriate classifier to make an accurate decision. There are many kinds of classifiers available in the literature, including support vector machine, k-nearest neighbor, neural networks, and autoencoders. These are among the most popular classifiers used in heart sound analysis [8, 10, 14, 21, 34, 35]. Nowadays, with the availability of powerful processing units, deep neural network classifiers such as convolutional neural networks, long short-term memory networks, and other types of deep networks have started to gain popularity [1, 15, 16, 30, 34–43].

This study mainly focuses in the feature extraction process and proposes a single time-varying spectral feature, termed as the *centroid frequency*, which enables the classifiers to make an accurate decision. The proposed method is tested for both binary and multiclass classification problems. The obtained results reveal the

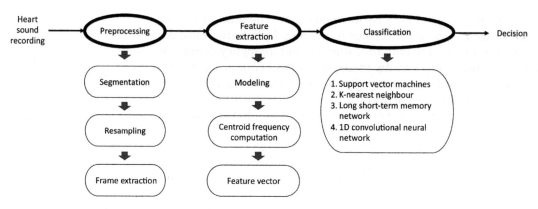

Fig. 1 Block diagram of the proposed method.

efficacy of the proposed method. The accuracy obtained for both cases are higher than the ones available in the literature despite the fact that only one kind of feature was used to feed the classifier.

2 Method

Fig. 1 illustrates the general overview of the method proposed in this study. There are three main blocks. These are the *preprocessing*, *feature extraction*, and *classification* blocks. These main blocks have their own subblocks to complete the main task. Firstly, a heart sound recording is fed into the preprocessing block, where it is segmented, resampled, and divided into frames. Secondly, the extracted frames are sent to the feature extraction block, where they are modeled, their features are extracted, and a feature vector is created. Finally, the obtained feature vector is given to a classification block as an input, where the selected classifier will provide the decision regarding the class of heart sound, whether it is normal or diseased.

The operation of all building blocks of the proposed scheme and the properties of the signals driving its certain blocks will now be explained in detail in the following sections. Special emphasis will be given to the feature extraction block, which constitutes the most critical steps in our proposed approach.

2.1 Database

The individual heart sound recordings used as the input to our proposed scheme are acquired from a well-developed open dataset [34]. The contents of the dataset are summarized in Table 1. The recordings in the database can be grouped into five classes.

Table 1 Details of the dataset.

Class	Subclass	No. of recordings	No. of cardiac cycles
Normal	Normal (N)	200	600
Abnormal	Aortic stenosis(AS)	200	600
	Mitral regurgitation (MR)	200	600
	Mitral stenosis (MS)	200	600
	Mitral valve prolapse (MVP)	200	600
Total		1000	3000

These are the *normal (N)*, *aortic stenosis (AS)*, *mitral regurgitation (MR)*, *mitral stenosis (MS)*, and *mitral valve prolapse (MVP)* classes. Each class contains 200 recordings, making 1000 recordings in total. Each recording contains a digital heart sound signal covering three cardiac cycles. The sampling frequency for the heart sound signal is the same for all recordings and is equal to 8 kHz. No filtering or standardization operation is performed on the data as these operations have already been performed by the original author of the database.

2.2 Preprocessing

Generally, the early step of preprocessing involves filtering and standardization or normalization of the signal, but in this case the data have already gone through these processes. Therefore, we now focus on the other steps to prepare data for subsequent feature extraction block.

2.2.1 Segmentation

There are two main assumptions in this study: one of them is that all the heart sound recordings in the database must have only one cardiac cycle. Therefore, with the aim of creating recordings containing single cycle, all the recordings were divided into three segments. This process was done manually. After this process, there were a total of 3000 heart sound recordings, for the original database contained 1000 recordings, with three cardiac cycles each. From here onward the term heart signal/cardiac signal will be used to refer a recording containing one cardiac cycle.

2.2.2 Resampling

The second important design constraint is that all the signals must have the same number of samples. In this case, all heart signal recordings are required to contain 1000 samples in them, irrespective of their unequal time duration. To meet this criterion the signals were first interpolated with factor 1000 and then decimated with a factor equaling their original signal length.

2.2.3 Frame extraction

After segmentation and resampling the cardiac signal was then divided into the frames of length 150 with a slide length 25. This created 35 frames from a single signal recording. The choice of frame size was made heuristically, and several experiments were performed to verify its efficacy. The reason for extracting frames was to obtain stationarity in frames of the phonocardiograms.

2.3 Feature extraction

2.3.1 Modeling

Two different methods, *discrete Fourier transform* (*DFT*) and *autoregressive* (*AR*) modeling, were used to model the given cardiac signals. These are among the most popular methods for spectrum estimation.

The discrete Fourier transform method

The DFT is a method to obtain a frequency domain representation of a finite sequence acquired by sampling a continuous function at regular intervals. It is widely used to investigate the spectral content of a finite length discrete-time signal sequence, which corresponds to a heart sound signal frame in this work. The mathematical expression for calculation of DFT is as follows:

$$X(k) = \sum_{n=0}^{N-1} x(n) e^{-j\frac{2\pi}{N}nk}$$

where n denotes the sample number of the signal samples within the heart sound signal frame, k denotes the frequency bin number of the discrete frequency-domain signal, and N denotes the total number of signal samples within the frame.

The autoregressive modeling method

The AR modeling method is an alternative to DFT-based spectral analysis. It is a well-developed and popular spectral analysis method because of its high spectral resolution and low spectral leakage [44]. An AR model of a finite length discrete-time signal sequence can be expressed in a difference equation form as follows:

$$x(n) = -\sum_{k=1}^{p} a_k x(n-k) + e(n)$$

Here, $x(n)$ denotes the signal to be modeled corresponding to the signal samples within the heart sound signal frame, a_k denote the AR model parameters, p denotes the model order, and $e(n)$ is a zero-mean white noise process.

It is easily seen from the previous equation that the AR method models the data as the output of a linear filter driven by a white noise process, which can be represented in z-domain as follows:

$$X(z) = H(z).E(z)$$

where $H(z)$ is the frequency response of the AR filter and $E(z)$ is the frequency response of the driving white process, which is actually a constant being equal to the variance of the process. The frequency response may be expressed in frequency domain as follows:

$$H(e^{j\Omega}) = \frac{X(e^{j\Omega})}{E(e^{j\Omega})} = \frac{1}{\left[1 + \sum_{k=1}^{p} a_k e^{-jk\Omega}\right]}$$

The frequency content of the signal $x(n)$ may be evaluated in a manner similar to the DFT by sampling the previous equation at equally spaced frequency bins.

2.3.2 Centroid frequency computation

Assuming that the frequency response of a frame is available, the centroid frequency for this frame can be calculated as follows:

$$f_c = \frac{\sum_{i=1}^{N} H(f_i) f_i}{\sum_{i=1}^{N} H(f_i)}$$

where N is the number of frequency points, f_i's denote the frequency points at which the frequency response is evaluated, and $H(f_i)$ denotes the value of the frequency response at frequency bin f_i.

The centroid frequency values calculated for successive frames of a heart sound signal covering one cardiac cycle constitute the elements of the resultant feature vector representing that cardiac cycle [45, 46]. Since we have generated 35 signal frames for each cardiac cycle, we obtained feature vectors of length 35 for each recording included in this work.

2.4 Classification

For classification the dataset is divided into two subsets: the *training* set (containing features and class labels) and the *testing* set (containing features only). The main task of the classifier algorithm is to train a predictive model using information from the training set, which will then be employed on the test set to predict their respective labels. Four different classifiers based on the methods of *k-nearest neighbor* ($k-NN$), *support vector machine* (*SVM*), *long short-time memory network* (*LSTM*), and *one-dimensional convolutional neural network* (*1-D-CNN*), have been implemented for classification process in this study. They are discussed briefly in the following section.

2.4.1 Support vector machine classifier

Support vector machine (SVM) is one of the most widely used methods for classification and utilizes machine learning theory to prevent overfitting of data to yield maximum accuracy for predicting the given task. SVM utilizes hypothetical space of linear functions to map the data into a high-dimensional feature space to classify the data that may not be separated linearly. The data are converted in a certain way that the divider could be fitted as a hyperplane in the space. A suitable divider is calculated; the divider is also known as the optimal hyperplane, which defines the boundaries of the classes. There are some data near to the optimal hyperplane, known as support vectors, which are very crucial in setting the boundaries.

2.4.2 k-nearest neighbor classifier

k-nearest neighbor (k-NN) is another popular classifier in data mining. It is usually chosen because of its simplicity and easiness to handle. In addition, the computational cost is low, and the obtained result is easily interpretable. The decision of the classifier depends on the distance from the neighbors. There are many types of functions that can be used to calculate the distance between the neighbors. In this case, we have used the Euclidean distance function to calculate the distance between neighbors.

2.4.3 One-dimensional convolutional neural network classifier

One-dimensional convolutional neural network (1-D-CNN) is a type of deep neural networks. Conventional 2-D CNNs are normally used in image classification problems and have different layers for training. The 1-D CNN has a similar architecture to 2-D CNNs, with the only difference being that they have one-dimensional inputs accepting data in the format of a time series. One-dimensional CNN has been found to perform well for shorter segments especially for audio signals.

Fig. 2 shows the structure of the proposed 1-D CNN classifier. The proposed 1-D CNN has a total of 15 layers: four convolution layers, four batch normalization layers, four rectified linear unit (ReLU) layers, two fully connected layers, and one softmax layer. In the convolution layers the convolution operation is performed between input vector and the kernels, and an output known as feature map containing features is created. The batch normalization and ReLU layers are often used to resolve the internal covariance deviation and to increase computation speed. The fully connected layers are similar to the hidden layers of neural networks. And the softmax layer contains softmax function employed to make decision for the multiclass classification tasks.

2.4.4 Long short-term memory network classifier

Long short-term memory network classifier (LSTM) is another type of deep neural networks. It is able to prevent the problem arising due to long-term dependency in time series. It has the property to memorize the data for longer periods compared with other deep neural network types, which helps to solve the

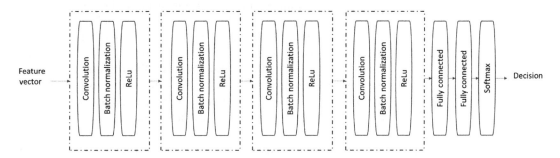

Fig. 2 Proposed 1-D CNN architecture.

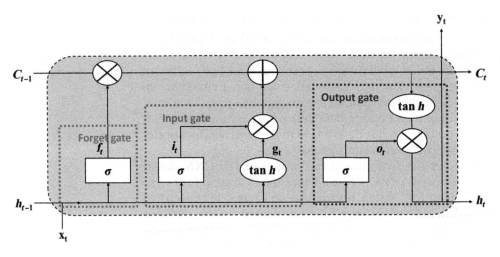

Fig. 3 LSTM network architecture.

sequence prediction problems. Since cardiac sound signal is a signal sequenced with respect to time, the LSTM network can be a better option.

The LSTM comprises four components: the *memory cell*, the *forget gate*, the *input gate*, and the *output gate*, which are illustrated in Fig. 3. The functions of the three gates are to help the memory cell in deciding what information should be saved, erased or updated during learning process, which ultimately reduces the data redundancy, and optimized the performance.

3 Performance evaluation

The performances of the methods discussed earlier are evaluated by adopting the fivefold cross-validation technique for validating performance metrics. The results for each classifier are finally expressed in the form of a confusion matrix for convenience. The information obtained from the confusion matrix is used in performance evaluation. The performances of the classifiers included in the classification experiments are measured by using four different metrics. These are *accuracy, sensitivity, specificity,* and *F1 score*.

For multiclass classification, these parameters should be calculated individually for each class using the same way as for binary classification and average of all classes should be taken at the end to obtain the overall values for those parameters.

4 Results and discussion

Both two-class and five-class classification tasks were performed in this study. These tasks were accomplished with the use of four different classifiers. All of these classifiers were simultaneously fed with two different feature vectors, to evaluate the effectiveness of the methods presented in this study for feature extraction. For the validation of performance metrics, fivefold cross validation method was opted. The obtained results are provided in the form of confusion matrices shown in Figs. 4 and 5.

The values of performance metrics for all the classifiers for both binary and multiclass classification were calculated and tabulated in Table 2.

During the binary classification the classifiers utilizing AR-based feature vector outperformed the classifiers utilizing FFT-based feature vector. Although the values of performance parameters were relatively similar, the fact that the AR feature vector fed classifiers obtained higher accuracies, sensitivities, specificities, and F1 scores cannot be neglected.

While making comparisons on the basis of classifier being used, for binary classification, k-NN classifier using the AR-derived feature vector achieved the highest accuracy of 99.60%. The second highest accuracy, 99.30%, was reached by SVM and 1-D CNN classifiers utilizing the AR feature vector. Similarly, when the

(A)

		SVM			
		FFT		AR	
		Predicted class		Predicted class	
		A	N	A	N
Actual class	A	2380	20	2392	8
	N	26	574	13	587

(B)

		k-NN			
		FFT		AR	
		Predicted class		Predicted class	
		A	N	A	N
Actual class	A	2377	23	2395	5
	N	16	584	7	593

(C)

		LSTM			
		FFT		AR	
		Predicted class		Predicted class	
		A	N	A	N
Actual class	A	2358	42	2379	21
	N	78	522	51	549

(D)

		1-D CNN			
		FFT		AR	
		Predicted class		Predicted class	
		A	N	A	N
Actual class	A	2380	20	2388	12
	N	30	570	9	591

Fig. 4 Confusion matrices obtained for binary classification using various classifiers.

226 Chapter 13 Accurate classification of heart sounds

(A)

		SVM									
		FFT					AR				
		Predicted class					Predicted class				
		AS	MR	MS	MVP	N	AS	MR	MS	MVP	N
Actual class	AS	560	15	10	14	1	568	10	9	12	
	MR	22	542	18	16	2	10	561	9	19	1
	MS	13	34	529	16	8	7	7	572	10	4
	MVP	19	21	18	535	7	8	18	4	567	3
	N	5	3	4	8	580	3	2	4	3	588

(B)

		k-NN									
		FFT					AR				
		Predicted class					Predicted class				
		AS	MR	MS	MVP	N	AS	MR	MS	MVP	N
Actual class	AS	572	12	6	8	2	572	9	9	9	1
	MR	18	541	21	16	4	4	572	9	13	2
	MS	24	24	531	13	8	3	2	587	5	3
	MVP	16	35	19	522	8	6	20	2	572	0
	N	5	1	7	5	582	3	1	3	1	592

(C)

		LSTM									
		FFT					AR				
		Predicted class					Predicted class				
		AS	MR	MS	MVP	N	AS	MR	MS	MVP	N
Actual class	AS	504	44	23	29	0	534	24	11	31	0
	MR	29	503	33	34	1	24	529	20	21	6
	MS	7	34	494	56	9	7	21	551	10	11
	MVP	26	60	41	470	3	16	26	12	541	5
	N	5	23	16	22	534	0	1	9	1	589

(D)

		1D CNN									
		FFT					AR				
		Predicted class					Predicted class				
		AS	MR	MS	MVP	N	AS	MR	MS	MVP	N
Actual class	AS	566	17	7	9	1	584	6	3	7	0
	MR	10	546	22	20	2	7	578	3	11	1
	MS	2	9	573	11	5	2	1	592	4	1
	MVP	12	21	14	549	4	6	7	0	587	0
	N	3	1	6	6	584	0	2	3	1	594

Fig. 5 Confusion matrices obtained for multiclass classification using various classifiers.

FFT-based feature vector was used as an input, the accuracies of k-NN, SVM, and 1-D CNN were 98.70%, 98.47%, and 98.33%, respectively. The worst performer among all classifiers was LSTM classifier with the least accuracy of 96.00% using the FFT feature vector. Even with the AR-based feature vector, it obtained only the accuracy of 97.60%.

Table 2 Comparison of the performances of the classifiers.

Classifier	Methods	Accuracy (%)		Sensitivity (%)		Specificity (%)		F1 score (%)	
		Binary	Multiclass	Binary	Multiclass	Binary	Multiclass	Binary	Multiclass
SVM	FFT	98.47	91.53	99.17	91.53	95.67	97.88	99.04	91.54
	AR	99.30	95.20	99.67	95.20	97.83	98.80	99.56	95.21
k-NN	FFT	98.70	91.60	99.04	91.60	97.33	97.90	99.19	91.60
	AR	**99.60**	96.50	**99.79**	96.50	**98.83**	99.12	**99.75**	96.50
LSTM	FFT	96.00	83.50	98.25	83.50	87.00	95.87	97.52	83.90
	AR	97.60	91.47	99.12	91.47	91.50	97.87	98.51	91.45
1-D-CNN	FFT	98.33	93.93	99.17	93.93	95.00	98.48	98.96	93.94
	AR	99.30	**97.84**	99.50	**97.84**	98.50	**99.46**	99.56	**97.84**

These bold values represent the highest value achieved by a particular classifier for a specific task.

The most interesting characteristics of the proposed method are that despite having unevenly distributed dataset (2400 abnormal cases and only 600 normal cases), the binary classification task was well executed by all the classifiers. In general, uneven distribution of data provides misleading accuracy rates. The classifiers classify the largely available cases accurately, while the groups with fewer cases are misclassified. But the accuracy rates in the result appear to be high because of the correct classification of dominant cases. Therefore, other parameters like specificity, sensitivity, and F1 score are required to be assessed. In our case, 98.83% specificity, 99.79% sensitivity, and 99.75% F1 score was obtained for the k-NN classifier whose accuracy was 99.60%. These parameters, especially the higher specificity, verified that the introduced feature extraction method is robust.

In the same way, for multiclass classification, the same classifiers were assessed for their performances. In this case the classifiers using AR-based feature vector outperformed the classifiers using FFT-based feature vector. However, the differences between the values of performance parameters of the different classifiers were more evident in the multiclass classification.

Further comparisons were made between the performances of the different classifiers using the same feature vector as an input. Using the AR-based feature vector, 1-D CNN classifier achieved the highest accuracy of 97.84%, whereas LSTM achieved the least accuracy of 91.47%. Similarly, an accuracy of 93.93% was obtained using FFT feature vector with 1-D CNN classifier, which is the highest among other classifiers utilizing the same feature vector.

On the other hand, the overall least accuracy, 91.47%, was attained using the FFT-derived feature vector with LSTM classifier. For both feature vectors, 1-D CNN performed the best, while the LSTM classifier performed the worst.

The variations in the resolution of spectrum obtained from different methods, that is, AR and FFT, used to extract the feature might be the reason for different levels of performances of same classifier with two different feature vectors. It is a well-known fact that the spectral resolution of the AR method is higher in comparison with different spectrum estimation methods available [47].

The LSTM and 1-D CNN classifiers were chosen in this work to study the efficacy of the state-of-the-art deep learning techniques in heart sound classification. Although the LSTM classifier did not perform as desired, the performance of 1-D CNN was excellent. The low performances of LSTM classifier might be due to the small dataset being used. In general, large datasets are used for training deep learning systems. But the ability of 1-D CNN to provide higher accuracy with a relatively small set of data is notable. This higher performance of deep learning methods in classification task comes with the price of long training times and larger memory requirements.

The main aim of this study was to introduce a single feature and be able to provide higher accuracy rates for cardiac sound classification. While comparing our results with others in the literature, it was found that the goal was reached. To verify aforementioned statement the details of the comparison between this study and another study conducted by Yaseen et al. [34] are summarized in Table 3.

Table 3 Comparison between this study and the study by Yaseen et al. [34].

S.N.			Accuracy	
			This study	Yaseen et al.
1.	Best result with single feature		97.84%	92.30%
2.	Best result with same classifiers (with single feature)	SVM	95.20%	92.30%
		K-NN	96.50%	91.80%
3.	Overall best result		97.84%	97.90%
			(1-D CNN with single feature)	(SVM with two features)

The numerical values in the bold showcase the highest accuracy achieved. The words in bold typeface represent the methods used by different authors.

The highest accuracy achieved by Yaseen et al. while using a single feature was 92.30%, which is less than the highest accuracy obtained in this study, 97.84%. Similarly, both studies have two classifiers in common: SVM and k-NN. In the work by Yaseen et al. [34], the SVM and k-NN classifiers fed with a single feature reached the maximum accuracy rates of 92.30% and 91.80%, respectively, whereas these same classifiers with single feature acquired the accuracy rates of 95.20% and 96.50% in our study. Although an overall highest accuracy, 97.90%, was obtained in the work of Yaseen et al. using the SVM classifier, this was only possible by combining two different features. On the other hand, an almost equal accuracy of 97.84% was obtained in this study by using 1-D CNN with a single feature. This also verifies the robustness of the feature presented in this study.

5 Conclusion

A time-varying spectral feature for cardiac sound classification, termed as centroid frequency, was introduced in this study. The idea of classifying cardiac sounds with the help of a single feature along with the method for extracting the desired feature was presented. Two different methods (FFT and AR) were tested. While comparing their performances the AR method was found to be extracting better spectral features than the FFT method. The classification results obtained with the AR-derived feature vector for both binary and multiclass classification were compared with the state-of-the-art techniques and found to be outstanding. Hence, it is clearly demonstrated that the proposed AR-based feature extraction method is capable of extracting time-varying spectral properties of the cardiac sounds in the form of a single time-varying feature and this feature can be used in heart sound analysis to yield very high accuracy rates.

Moreover, the accuracy of 97.84% for multiclass classification obtained using 1-D CNN illustrates the success of deep learning in the classification of cardiac sounds. Therefore, the possibility of the use of deep learning-based classifiers in the heart disease diagnosis in the near future can be clearly indicated. The success achieved in this study signals the possible use of this proposed method in the analysis and classification of other human physiological signals, which are nonstationary in nature.

References

[1] O. Faust, Y. Hagiwara, J.H. Tan, S.L. Oh, U.R. Acharya, Deep learning for healthcare applications based on physiological signals: a review, Computer Methods and Programs in Biomedicine 161 (2018) 1–13.

[2] World Health Organization, 2018. The Top 10 Causes of Death. (Web page: https://www.who.int/news-room/fact-sheets/detail/the-top-10-causes-of-death).
[3] Ritchie, H., 2019. What Do the People of the World Die From?. (web page: https://www.bbc.com/news/health-47371078).
[4] World Health Organization, 2017. Cardiovascular Diseases (CVDs). (Web page: https://www.who.int/en/news-room/fact-sheets/detail/cardiovascular-diseases-(cvds)).
[5] M.R. Montinari, S. Minelli, The first 200 years of cardiac auscultation and future perspectives, Journal of Multidisciplinary Healthcare 12 (2019) 183–189.
[6] M.A. Chizner, Cardiac auscultation: rediscovering the lost art, Current Problems in Cardiology 33 (7) (2008) 326–408.
[7] M.N. Anas, M.F. Shadi, The heart auscultation: from sound to graphical, ARPN J. Eng. Appl. Sci. 9 (2014) 1924–1929.
[8] H. Uguz, A biomedical system based on artificial neural network and principal component analysis for diagnosis of the heart valve diseases, Journal of Medical Systems 36 (1) (2012) 61–72.
[9] M. Singh, A. Cheema, Heart sounds classification using feature extraction of phonocardiography signal, International Journal of Computers and Applications 77 (2013) 13–17.
[10] H. Tang, Z. Dai, Y. Jiang, T. Li, C. Liu, PCG classification using multidomain features and SVM classifier, BioMed Research International 2018 (2018) 4205027.
[11] E.F. Gomes, P.J. Bentley, E. Pereira, M.T. Coimbra, Y. Deng, Classifying heart sounds—approaches to the PASCAL challenge, in: Proceedings of the International Conference on Health Informatics (HEALTHINF-2013), 2013, pp. 337–340.
[12] Z. Sharif, M.S. Zainal, A.Z. Sha'ameri, S.H.S. Salleh, Analysis and classification of heart sounds and murmurs based on the instantaneous energy and frequency estimations, in: TENCON, IEEE Region 10 International Conference, September 24–27, 2000, Malaysia, Kuala Lumpur, 2000, pp. 130–134.
[13] J.K. Roy, T.S. Roy, S.C. Mukhopadhyay, Heart sound: detection and analytical approach towards diseases, in: S. Mukhopadhyay, K. Jayasundera, O. Postolache (Eds.), Modern Sensing Technologies, Springer, Germany, 2019, pp. 103–145.
[14] Y. Koçyiğit, Heart sound signal classification using fast independent component analysis, Turkish Journal of Electrical Engineering and Computer Sciences 24 (4) (2016) 2949–2960.
[15] B. Bozkurt, I. Germanakis, Y. Stylianou, A study of time-frequency features for CNN-based automatic heart sound classification for pathology detection, Computers in Biology and Medicine 100 (2018) 132–143.
[16] F. Noman, C. Ting, S. Salleh, H. Ombao, Short-segment heart sound classification using an ensemble of deep convolutional neural networks, in: ICASSP 2019–2019 IEEE International Conference on Acoustics, Speech and Signal Processing (ICASSP), May 12–17, 2019, Brighton, United Kingdom, 2019, pp. 1318–1322.
[17] S. Yuenyong, A. Nishihara, W. Kongprawechnon, K. Tungpimolrut, A framework for automatic heart sound analysis without segmentation, Biomedical Engineering Online 10 (13) (2011) 1–23.
[18] A. Moukadem, A. Dieterlen, N. Hueber, C. Brandt, A robust heart sounds segmentation module based on S-transform, Biomedical Signal Processing and Control 8 (3) (2013) 273–281.
[19] Y. Soeta, Y. Bito, Detection of features of prosthetic cardiac valve sound by spectrogram analysis, Applied Acoustics 89 (2015) 28–33.

[20] H. Nazeran, Wavelet-based segmentation and feature extraction of heart sounds for intelligent PDA-based phonocardiography, Methods of Information in Medicine 46 (2007) 135–141.
[21] W. Kao, C. Wei, J. Liu, P. Hsiao, Automatic heart sound analysis with short-time Fourier transform and support vector machines, in: 52nd IEEE International Midwest Symposium on Circuits and Systems, August 2–5, 2009, Mexico, Cancun, 2009, pp. 188–191.
[22] P.S. Vikhe, N.S. Nehe, V.R. Thool, Heart sound abnormality detection using short time fourier transform and continuous wavelet transform, in: Second International Conference on Emerging Trends in Engineering and Technology, ICETET-09, December 16–18, 2009, India, Maharashtra, 2009, pp. 50–54.
[23] V. Millette, N. Baddour, Signal processing of heart signals for the quantification of non-deterministic events, Biomedical Engineering Online 10 (10) (2011).
[24] F. Nogata, Y. Yokota, Y. Kawanura, H. Morita, Y. Uno, W.R. Walsh, Audio-visual based recognition of auscultatory heart sounds with Fourier and wavelet analyses, Global J. Technol. Optimization 3 (2012) 42–48.
[25] N. Kehtarnavaz, Digital Signal Processing System Design (Second Edition), Academic Press, United States, 2008 334 pp.
[26] O. Rioul, P. Flandrin, Time-scale energy distributions: a general class extending wavelet transforms, IEEE Transactions on Signal Processing 40 (7) (1992) 1746–1757.
[27] M.E. Karar, S.H. El-Khafif, M.A. El-Brawany, Automated diagnosis of heart sounds using rule-based classification tree, Journal of Medical Systems 41 (2017) 60.
[28] S.M. Debbal, F. Bereksi-Reguig, Detection of differences of the phonocardiogram signals by using the continuous wavelet transform method, Int. J. Biomedical Soft Computing Hum. Sci. 18 (2) (2013) 73–81.
[29] S.M. Debbal, A.M. Tani, Heart sounds analysis and murmurs, International Journal of Medical Engineering and Informatics 8 (1) (2016) 49–62.
[30] A. Meintjes, A. Lowe, M. Legget, Fundamental heart sound classification using the continuous wavelet transform and convolutional neural networks, in: 40th Annual International Conference of the IEEE Engineering in Medicine and Biology Society (EMBC), July 17–21, 2018, Honolulu, 2018, pp. 409–412.
[31] O. Yıldız, A. Arslan, Automated auscultative diagnosis system for evaluation of phonocardiogram signals associated with heart murmur diseases, **Gazi** Univ. J. Sci. 31 (1) (2018) 112–124.
[32] J.J.G. Ortiz, C.P. Phoo, J. Wiens, Heart sound classification based on temporal alignment techniques, in: 2016 Computing in Cardiology Conference (CinC), September 11–14, 2016, Vancouver, Canada, 2016, pp. 589–592.
[33] M. Hamidi, H. Ghassemian, M. Imani, Classification of heart sound signal using curve fitting and fractal dimension, Biomed. Signal Process. Control 39 (2018) 351–359.
[34] Yaseen, S. Gui-Young, K. Sooni, Classification of heart sound signal using multiple features, Appl. Sci. 8 (12) (2018) 2344.
[35] A.K. Dwivedi, S.A. Imtiaz, E. Rodriguez-Villegas, Algorithms for automatic analysis and classification of heart sounds - a systematic review, IEEE Access 7 (2018) 8316–8345.
[36] A. Keleş, Derin öğrenme ve sağlık alanındaki uygulamaları, Turkish Studies-Information Technologies and Applied Sciences 13 (21) (2018) 113–127.
[37] J.X. Low, K.W. Choo, Classification of heart sounds using softmax regression and convolutional neural network, in: Proceedings of the 2018 International Conference on Communication Engineering and Technology, ACM: 18–21, 2018.
[38] P. Samui, S.S. Roy, V.E. Balas (Eds.), Handbook of Neural Computation, Academic Press, 2017.

[39] S. Latif, M. Usman, R. Rana, J. Qadir, Phonocardiographic sensing using deep learning for abnormal heartbeat detection, IEEE Sens. J. 18 (22) (2018) 9393–9400.
[40] V.E. Balas, S.S. Roy, D. Sharma, P. Samui (Eds.), Handbook of Deep Learning Applications, In: vol. 136, Springer, 2019.
[41] R. Biswas, A. Vasan, S.S. Roy, Dilated Deep Neural Network for Segmentation of Retinal Blood Vessels in Fundus Images, in: Transactions of Electrical Engineering, Iranian Journal of Science and Technology2019, pp. 1–14.
[42] E. Messner, M. Zöhrer, F. Pernkopf, Heart sound segmentation - an event detection approach using deep recurrent neural networks, IEEE Trans. Biomed. Eng. 65 (9) (2018) 1964–1974.
[43] T.E. Chen, S.I. Yang, L.T. Ho, K.H. Tsai, Y.H. Chen, Y.F. Chang, C.C. Wu, S1 and S2 heart sound recognition using deep neural networks, IEEE Trans. Biomed. Eng. 64 (2) (2016) 372–380.
[44] R. Takalo, H. Hytti, H. Ihalainen, Tutorial on univariate autoregressive spectral analysis, J. Clin. Monit. Comput. 19 (2005) 401–410.
[45] P. Upretee, M.E. Yüksel, Accurate classification of heart sounds for disease diagnosis by a single time-varying spectral feature: preliminary results, in: Proceedings of 2019 Scientific Meeting on Electrical-Electronics & Biomedical Engineering and Computer Science (EBBT), Istanbul, Turkey, 2019.
[46] P. Upretee, Classification of Heart Sounds for Disease Diagnosis By Using a Single Time-Varying Spectral Feature, M.S. thesis (in Turkish)Erciyes University, Kayseri, Turkey, 2019.
[47] Ö. Deperlioğlu, Classification of segmented heart sounds with artificial neural networks, International Journal of Applied Mathematics Electronics and Computers 6 (4) (2018) 39–44.

Complex neutrosophic δ-equal concepts and their applications in water quality

Prem Kumar Singh
Amity School of Engineering and Technology, Amity University, Noida, Uttar Pradesh, India

1 Introduction

Nowadays, many researchers are focusing their attention on precise measurement of indeterminacy using three-way decision spaces. The three-way space gives a precise way to characterize the uncertainty, inconsistency, and incompleteness of fuzzy attributes based on truth, indeterminacy, and falsity membership values [1–7]. Several other methods are also introduced for handling uncertainty in the three-way space, including shadow sets [8, 9], interval-valued sets [10, 11], and neutrosophic sets [2, 12, 13]. This chapter focuses on the analysis of three-way fuzzy attributes and their dynamic changes using the properties of neutrosophic sets [1, 2, 14–16]. One of the major problems with three-way fuzzy set is that it unable to measure the changes in uncertain attributes [16]. The precise representation of uncertainty and its fluctuation in three-way fuzzy attributes is a major task for data analytics researchers. It is indeed a requirement for many organizations and industries [2]. For example, the opinions of people toward conflicts among India and Pakistan, Philistine and Israel, the United States and Russia, and North and South Korea can and will change several times. The reason for this is that while some countries may support India, other countries may support Pakistan, and still other countries are neutral or refrain from interfering in general. Numerically, suppose six countries voted in support of resolution, three countries voted against resolution, and one country is uncertain or neutral about the conflict in a given year. This three-way cognitive context can be written via truth, indeterminacy, and falsity membership values of a neutrosophic set (0.6, 0.1, 0.3), independently. In case the opinion of countries changes in the next year based on some given parameters, then analysis of fluctuation existing in the attributes becomes computationally expensive. This creates a major issue for any political or data scientist of a democratic country in analyzing the opinion of people toward win, loss, and neutrality of a given political party. Our daily life utility like air and water pollution changes each phase of time. In this case, it will be difficult for the concerned manufacturing company to measure the fluctuation and produce the particular tool. Another problem arises when an expert wants to measure the effect of human thought on the manufacturing company based on its closest distance or defined granulation. To conquer these problems, we introduce a complex neutrosophic context and its graphical structure visualization and navigation at different granulation. The objective is to provide a mathematical model for graphical analytics of complex neutrosophic context in a more descriptive way when compared to its numerical representation. The motivation is to extract some of the meaningful δ-equal complex neutrosophic concepts from the given complex neutrosophic context at user-defined complex granules for knowledge processing tasks as shown in Fig. 1. To understand this necessity, we present a list of research papers in Table 1. The $*$ mark in Table 1 shows that less attention has been paid toward analysis of complex neutrosophic attributes, and more research is needed. To fill this gap, this chapter focuses on introducing a basic algorithm for adequate analysis of three-way complex fuzzy attributes using the calculus of applied abstract algebra and lattice theory.

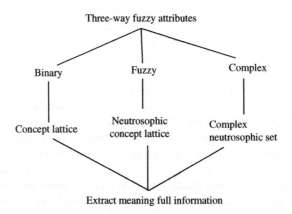

Fig. 1 A graphical representation of three-way fuzzy attributes.

Table 1 List of research papers to understand the necessity of the research gap.

	Bipolar fuzzy space	Three-way fuzzy space	Complex fuzzy space	Three-way complex space
Formal context	[2, 14]	[12, 17]	[18–21]	*, [16]
Formal concept	[22, 23, 25, 26]	[1, 3, 24, 27]	[16]	*
Concept lattice	[14, 15, 29]	[1–3, 6, 7, 30]	[16, 28]	*
Attribute implication	[14, 17, 32]	[12, 13]	[31]	*
Distance measurement	[2, 15]	[33, 34]	[35]	[36]
Granular computing	[9, 37, 38]	[3, 39]	[40, 41]	*
Concept learning	[24, 38]	[4, 42]	[35, 43]	*, [36]

Complex neutrosophic attributes can be found in many periodic datasets, including those on health care [15], air quality [2], the stock market [13, 44], as well as time series [10, 11, 40, 45] datasets. These are nothing but extensions of fuzzy sets [46–48] in complex fields [21, 41, 49] for handling dynamic attributes [50, 51] as shown in Table 2. The precise interpretation of these attributes is a major task due to the fact that user preferences change over time. For example, the meaning of temperature changes several times in a given year: 20°C temperature used to be considered warm in the winter season and cool in the summer season, but neither warm nor cool (i.e., indeterminant or uncertain) in the spring season. In a similar manner, the saturation value (intensity and brightness) and hue of colors change at each phase of time based on the combination of red-green-blue (RGB).[a] The precise mathematical representation of any given three-way complex or dynamic fuzzy attributes is computationally expensive. To deal with this, properties of the complex neutrosophic set [36] and its graphical representation were introduced [52] by some researchers for analysis of multidecision processes [3, 16]. The complex neutrosophic set is nothing but a generalization of neutrosophic sets where a new phase term is added to characterize the changes in truth, indeterminacy, and falsity membership values, independently. Hence, it can be considered as one of the hybridizations among complex

---

[a] See https://en.wikipedia.org/wiki/Complex_logarithm.

Table 2 Analysis of different extensions of fuzzy sets based on various parameters.

	Fuzzy set	Complex fuzzy set	Neutrosophic set	Complex neutrosophic set
Domain	Universe of discourse	Universe of discourse	Universe of discourse	Universe of discourse
Codomain	Single-valued in [0, 1]	Single-valued in unit circle [0, 1]	Three-valued in $[0, 1]^3$	Three-valued in unit circle $[0, 1]^3$
Uncertainty measurement	Yes	Yes	Yes	Yes
Unit circle	No	Yes	No	Yes
Truth membership	Yes in [0, 1]	Yes in [0, 1]	Yes in $[0, 1]^3$	Yes in $[0, 1]^3$
Indeterminacy membership	No	No	Yes in $[0, 1]^3$	Yes in $[0, 1]^3$
False membership	Negation of membership	Negation of membership	Yes in $[0, 1]^3$	Yes in $[0, 1]^3$
Amplitude term	No	Yes in $[0, 1]^3$	No	Yes in $[0, 1]^3$
Phase term measurement	No	Yes in $[0, 2\pi]$	No	Yes in $[0, 2\pi]$

fuzzy sets and neutrosophic sets. It is well known that the complex fuzzy sets provide a mathematical representation of those periodic problems that are unsolvable in one-dimensional single-valued membership values. It gives a way to represent the membership values of any attributes $\mu_Z(z)$ within the unit circle of a complex plane in the form $\mu_Z(z) = r_z(x)e^{iw_z(x)}$, where $i = \sqrt{-1}$, both $r_Z(z)$ and $w_Z(z)$ are real-valued, and $r_Z(z) \in [0, 1]$. In this set the amplitude [0, 1] represents uncertainty in the given attributes, whereas the fluctuation appears in the phase term $[0, 2\pi]$. The same mathematical properties of the complex fuzzy set are utilized in this chapter to represent the fluctuation in truth, indeterminacy, and falsity membership values of a defined single-valued neutrosophic set. This hybridization is called a complex neutrosophic set as discussed in [36]. However, the chapter focuses on deriving some useful information from the given complex neutrosophic datasets for knowledge processing tasks using the calculus of applied lattice theory [16] and complex neutrosophic graphs [52].

It can be observed that the complex neutrosophic set provides a way to characterize uncertainty and its changes based on a defined truth amplitude and its phase term, and indeterminacy amplitude and its phase term, falsity amplitude and its phase term, independently for precise representation of human thoughts and their changes at given phases of time. The complex neutrosophic graph [16, 52] provides a way to visualize them as vertices and nodes of a defined graph for better understanding. However, these papers are unable to provide some of the meaningful information of the given contexts when compared to the neutrosophic concept lattice [1, 2]. Sometime the user wants to know the hidden pattern in complex neutrosophic datasets for knowledge processing tasks. In this case, all of the available approaches lack a way to provide a mathematical way to deal with complex neutrosophic datasets. To achieve this goal, the chapter focuses on utilizing the algebra of a complex neutrosophic set [36] and its graphical representation [52] for the concept lattice [16] representation using the calculus of next neighbor and δ-equal complex granulation to zoom in and zoom out the complex neutrosophic context at different granulations. This granular computing provides a micro and macro level to characterize the complex dataset by multilevel views to understand the given problem, which captures the particular pattern in datasets. The motivation is to characterize the uncertainty and fluctuation in the complex fuzzy attributes based on truth amplitude and its phase term, indeterminacy amplitude and its phase term, and falsity amplitude and its phase term, independently, for qualitative view of knowledge processing tasks and the ensuing process of granular computing. The objective is to provide a compact visualization of complex neutrosophic contexts using the vertices and edges of a defined complex neutrosophic graph and lattice theory in such a way that complex granules provide a more intuitive and descriptive way to deal with complex neutrosophic contexts in various fields for knowledge discovery and representation tasks. In this way, the proposed mathematical tool is a more readable and for the complex neutrosophic contexts when compared to any of the available approaches for knowledge processing tasks. Some of the significant contributions of this chapter are as follows:

(i) The proposed method introduces a way to characterize the uncertainty and fluctuation in complex fuzzy attributes based on truth amplitude and phase term, indeterminacy amplitude and phase term, and falsity amplitude and phase term, independently.

(ii) The proposed method introduces a mathematical way for the concept lattice theoretical approach to define the super and subconcept hierarchical ordering among the complex neutrosophic datasets for precise analysis of knowledge processing tasks.

(iii) The proposed method introduces a method for processing the three-way complex neutrosophic context based on next-neighbor and user-defined complex granulation.
(iv) The proposed method introduces a method for investigating the δ-equal complex neutrosophic concepts with application for water quality measurement.
(v) The knowledge extracted from each of the proposed methods is also compared to validate the results with their advantages and disadvantages.
(vi) The proposed method will be helpful for water cleaning and other applications in other industries where changes in attributes exist at each phase of time.

The remainder of this chapter is structured as follows: Section 2 provides preliminaries for three-way complex neutrosophic graph visualization. Section 3 contains the proposed methods with their step-by-step illustration in Section 4. Section 5 concludes.

2 Three-way complex neutrosophic set and its graph

In this section, we discuss basic preliminaries of three-way formal concepts and their connection with complex neutrosophic sets are:

Definition 1 (Complex fuzzy set [19]) A complex fuzzy set Z can be defined over a universe of discourse U having a single fuzzy membership value at given phase of time. The complex-valued grade of membership of an element $z \in U$ can be characterized by $\mu_Z(z)$. The membership values that $\mu_Z(z)$ may receive all lie within the unit circle in the complex plane in the form $\mu_Z(z) = r_z(x)e^{iw_z(x)}$, where i=$\sqrt{-1}$, both $r_Z(z)$ and $w_Z(z)$ are real-valued, and $r_Z(z) \in [0, 1]$. The complex fuzzy set Z may be represented as a set of ordered pairs:

$$Z = \{(z, \mu_Z(z)) : z \in U\} = \{(z, r_Z(z)e^{iw_Z(z)}) : z \in U\}$$

Example 1. Let us suppose an expert wants to measure the pH level of water in a particular area.[b] It is well known that pH level of water changes several times. The expert observed a 50% pH level increase in the particular area at 6–7 months. The precise representation of this type of dataset is a major issue for the research community. It can be written using the properties of a complex fuzzy set where x_1 is area and y_1 is pH level. The changes in pH level can be written using the amplitude term [0, 1], whereas the time via phase term can be written as 2π. This complex word problem can be represented using

---

[b] See https://en.wikipedia.org/wiki/PH.

the properties of complex fuzzy sets as follows: $x_1 = 0.5e^{1.2\pi}/y_1$. Similarly, the opinion of more than two experts can be analyzed using the union and intersection of complex fuzzy sets $\mu_{z_1}(z) = r_{z_1}(z) \cdot e^{i\arg_{z_1}(z)}$ and $\mu_{z_2}(z) = r_{z_2}(z) \cdot e^{i\arg_{z_2}(z)}$ as given below [20]:

- $\mu_{z_1 \cup z_2} = r_{z_1 \cup z_2}(z) \cdot e^{i\arg_{z_1 \cup z_2}(z)} = \max(r_{z_1}(z), r_{z_2}(z)) \cdot e^{i\max(\arg_{z_1}(z), \arg_{z_2}(z))}$,
  $\mu_{z_1 \cap z_2} = r_{z_1 \cap z_2}(z) \cdot e^{i\arg_{z_1 \cap z_2}(z)} = \min(r_{z_1}(z), r_{z_2}(z)) \cdot e^{i\max(\arg_{z_1}(z), \arg_{z_2}(z))}$.

Definition 2 (Three-way fuzzy context [1]). This represents a formal context $\mathbf{K} = (X, Y, \widetilde{R})$ having set of objects X, set of attributes Y, and a neutrosophic relationship among them $\widetilde{R}$. It can be represented via an L-relation among X and Y, $\widetilde{R}: X \times Y \to L$ where L set can be defined independently in $[0, 1]$ for representing the truth, indeterminacy, and false membership values. It means three-way fuzzy attributes $y \in Y$ can be characterized by a truth membership function $T_N(y)$, an indeterminacy membership function $I_N(y)$, and a falsity membership function $F_N(y)$. The $T_N(y)$, $I_N(y)$, and $F_N(y)$ are real standard or nonstandard subsets of $]0^-, 1^+[$ as given below:

$$T_N : Y \to]0^-, 1^+[,$$
$$I_N : Y \to]0^-, 1^+[,$$
$$F_N : Y \to]0^-, 1^+[.$$

The neutrosophic set can be represented as follows:

$$N = \{(x, T_N(y), I_N(x), F_N(y)) : y \in Y\} \text{ where } 0^- \leq T_N(y) + I_N(y) + F_N(y) \leq 3^+.$$

It is noted that $0^- = 0 - \epsilon$ where 0 is its standard part and ϵ is its nonstandard part. Similarly, $1^+ = 1 + \epsilon$ ($3^+ = 3 + \epsilon$) where 1 (or 3) is its standard part and ϵ is its nonstandard part. The real standard (0, 1) or [0, 1] can also be used to represent the neutrosophic set. The union and intersection among neutrosophic sets N_1 and N_2 can be computed as follows [12, 53]:

$$N_1 \bigcup N_2 = \{(x, T_{N_1}(x) \vee T_{N_2}(x), I_{N_1}(x) \wedge I_{N_2}(x), F_{N_1}(x) \wedge F_{N_2}(x)) : x \in X\}.$$

The intersection of N_1 and N_2 can be defined as follows:

$$N_1 \bigcap N_2 = \{(x, T_{N_1}(x) \wedge T_{N_2}(x), I_{N_1}(x) \vee I_{N_2}(x), F_{N_1}(x) \vee F_{N_2}(x)) : x \in X\}.$$

This helps in finding a supremum and an infimum among three-way formal fuzzy concepts and their hierarchical visualization the concept lattice.

Example 2. Let us suppose that an expert wants to write the acceptation, rejection, and uncertain levels of pH (y_1) in a particular area (x_1). In this case, the expert can write using the properties of a neutrosophic set where truth value represents the acceptation level, false value represents the rejection level, and indeterminacy values are uncertain. For example, (0.5, 0.3, 0.2) represents that the pH level is 50% accepted, 30% rejected, and 20% uncertain about the water quality of a given area. However, this set is unable to represent the fluctuation or changes in the opinion of the expert at a given phase of time. To deal with this problem, the definition of complex neutrosophic set can be utilized.

Definition 3 (Three-way complex neutrosophic set [36]). A complex neutrosophic set Z can be defined over a universe of discourse U. The uncertainty in the attributes $z \in U$ can be characterized by true membership value $^-0 < r_{T_z} < 1^+$, indeterminacy membership value $^-0 \leq r_{I_z} < 1^+$, and falsity membership value $^-0 \leq r_{F_z} < 1^+$, independently, with a given phase of time $(0, 2\pi)$. It can be observed that, the "amplitude" term in the complex neutrosophic set satisfies the property $^-0 \leq r_{T_z} + r_{I_z} + r_{F_z} \leq 3^+$, whereas the "phase" term can be characterized by $w_{T_z}^r$, $w_{I_z}^r$, and $w_{F_z}^r$ in real-valued interval $[0, 2\pi]$. It can be represented as $Z = \left\{ (z, (r_{T_z} e^{w_{T_z}^r}, r_{I_z} e^{w_{I_z}^r}, r_{F_z} e^{w_{F_z}^r})) : z \in U \right\}$.

Example 3. Let us extend Example 2. The expert concluded that pH level of water in the particular area (x_1) is 50% accepted in 6–7 months, 30% rejected in 9–10 months, and 20% uncertain at end of a given year. This type of three-way complex linguistics words can be characterized by truth, indeterminate, and falsity membership values of a defined complex neutrosophic set as follows:

$$x_1 = (0.5e^{i1.2\pi}, 0.3e^{i1.8\pi}, 0.2e^{i2\pi})/y_1$$

where 2π is considered as phase term to represent any year. This three-way complex fuzzy attributes dataset can be visualized in compact format using the calculus of neutrosophic graph and its extensive properties.

Definition 4 (Three-way complete neutrosophic graph [54]). $G = (V, E)$ is a neutrosophic graph vertices V characterized by a truth membership function $T_V(v_i)$, an indeterminacy membership function $I_V(v_i)$, and a falsity membership function $F_V(v_i)$ as given below:

$$\left\{ (T_V(v_i), I_V(v_i), F_V(v_i)) \in [0, 1]^3 \right\} \text{ for all } v_i \in V.$$

Similarly, the edges $E\left\{ (T_E(V \times V), I_E(V \times V), F_E(V \times V)) \in [0, 1]^3 \right\}$ for all $(v_i v_j) \in V$ can be characterized via truth, falsity, and indeterminacy membership values as given below:

$$T_E(v_iv_j) = \min[T_E(v_i, T_E)(v_j)],$$
$$I_E(v_iv_j) = \max[I_E(v_i, I_E)(v_j)],$$
$$F_E(v_iv_j) = \max[F_E(v_i, F_E)(v_j)].$$

It is noted that $\{T_E(v_iv_j), I_E(v_iv_j), F_E(v_iv_j)\} = (0,0,0) \forall (v_i, v_i) \in (V \times V \backslash E)$.

Example 4. Let us extend Example 3, as the expert provides opinion about acceptation, rejection, and uncertainty levels of water pH for more than three regions. In this case, it can be visualized using the vertices and edges of a neutrosophic graph. Table 3 represents expert opinions about acceptation, rejection, and uncertain levels of pH in water of regions v_1, v_2, and v_3, whereas the corresponding relationship among them is shown in Table 4. This three-way information can be visualized using the vertices (V) and edges (E) of a complete neutrosophic graph as shown in Fig. 2.

Definition 5 (Complex neutrosophic graph [16, 52]). A complex neutrosophic fuzzy graph $G = (V, \mu_c, \rho_c)$ is a nonempty set in which the value of vertices $\mu_c : V \to (r_c^T(v) \cdot e^{i\arg_c^T(v)}, r_c^I(v) \cdot e^{i\arg_c^I(v)},$

Table 3 Expert opinion of water pH levels in the given areas for Example 4.

	v_1	v_2	v_3
T_V	0.8	0.6	0.2
I_V	0.2	0.3	0.1
F_V	0.1	0.1	0.8

Table 4 A neutrosophic relationship among pH level of the three regions in Example 4.

	v_1v_2	v_2v_3	v_3v_1
T_E	0.6	0.2	0.2
I_E	0.3	0.3	0.2
F_E	0.1	0.8	0.8

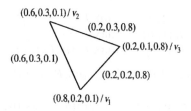

Fig. 2 A three-way neutrosophic graph representation of Tables 3 and 4.

$r_c^F(v) \cdot e^{i\arg_c^F(v)})$ and edges $\rho_c: V \times V \to (r_c^T(v \times v) \cdot e^{i\arg_c^T(v \times v)}, r_c^I(v \times v) \cdot e^{i\arg_c^I(v \times v)}, r_c^F(v \times v) \cdot e^{i\arg_c^F(v \times v)})$. It means the membership values can be characterized by truth, indeterminate, and falsity membership values within the unit circle of a complex argand plane in the given period of time. It can be represented through amplitude and phase term of the defined complex neutrosophic set as follows:

$$r_c^T(v_i \times v_j) \cdot e^{i\arg_c^T(v_i \times v_j)} \leq \min\left(r_c^T(v_i), r_c^T(v_i)\right) \cdot e^{i\min\left(\arg_c^T(v_i), \arg_c^T(v_j)\right)},$$
$$r_c^I(v_i \times v_j) \cdot e^{i\arg_c^I(v_i \times v_j)} \geq \max\left(r_c^I(v_i), r_c^I(v_i)\right) \cdot e^{i\max\left(\arg_c^I(v_i), \arg_c^I(v_j)\right)},$$
$$r_c^F(v_i \times v_j) \cdot e^{i\arg_c^F(v_i \times v_j)} \geq \max\left(r_c^F(v_i), r_c^F(v_i)\right) \cdot e^{i\max\left(\arg_c^F(v_i), \arg_c^F(v_j)\right)}.$$

The given complex fuzzy graph is complete iff:

$$r_c(v_i \times v_j) \cdot e^{i\arg_c(v_i \times v_j)} = \min\left(r_c(v_i), r_c(v_i)\right) \cdot e^{i\min\left(\arg_c(v_i), \arg_c(v_j)\right)}$$

for the truth, indeterminacy, and falsity membership functions, independently.

Example 5. Let us combine Examples 3 and 4. This provides a fluctuation in acceptation, rejection, and uncertain levels of water pH for the particular areas at given phases of time. It can be represented using the complex neutrosophic set as shown in Table 5. Similarly, the corresponding complex neutrosophic relationships

Table 5 A complex neutrosophic set representation of expert opinions about pH level for Example 5.

Vertex	y_1
x_1	$(0.5e^{j0.7\pi}, 0.3e^{j1.2\pi}, 0.2e^{j1.8\pi})$
x_2	$(0.7e^{j0.2\pi}, 0.6e^{j1.6\pi}, 0.1e^{j0.4\pi})$
x_3	$(0.4e^{j0.4\pi}, 0.5e^{j0.8\pi}, 0.6e^{j2\pi})$
x_4	$(0.8e^{j0.3\pi}, 0.7e^{j1.7\pi}, 0.3e^{j0.7\pi})$

among them can be represented using the edges as shown in Table 6. These two contexts can be visualized in a compact format using the vertices V and edges E of a defined complex neutrosophic graph as shown in Fig. 3. This represents unable to solve the problem of healthcare industry in case of suitability of water quality. To solve this problem, we introduce some new methods in the next section.

Definition 6 (Lattice structure of neutrosophic set [12, 13]). Let N_1 and N_2 be two neutrosophic sets in the universe of discourse X. Then, $N_1 \subseteq N_2$ iff $T_{N_1}(x) \leq T_{N_2}(x)$, $I_{N_1}(x) \geq I_{N_2}(x)$, $F_{N_1}(x) \geq F_{N_2}(x)$ for any $x \in X$. $(N, \wedge, \vee)$ is a bounded lattice. Also the structure $(N, \wedge, \vee, (1, 0, 0), (0, 1, 1), \neg)$ follows De Morgan algebra. Similarly, this lattice structure can be used to represent the three-way fuzzy concept lattice using Godel logic.

Table 6 A complex neutrosophic relation among each level of each region for Example 5.

Edges	y_1
$\{x_1, x_2\}$	$(0.5e^{i0.2\pi}, 0.3e^{i1.2\pi}, 0.1e^{i0.4\pi})$
$\{x_1, x_3\}$	$(0.4e^{i0.4\pi}, 0.3e^{i0.8\pi}, 0.2e^{i1.8\pi})$
$\{x_2, x_4\}$	$(0.7e^{i0.2\pi}, 0.6e^{i1.6\pi}, 0.1e^{i0.4\pi})$
$\{x_3, x_4\}$	$(0.4e^{i0.3\pi}, 0.5e^{i0.8\pi}, 0.3e^{i0.7\pi})$

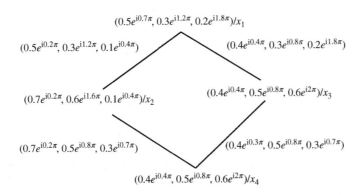

Fig. 3 A three-way complex neutrosophic graph visualization of context shown in Tables 5 and 6.

Definition 7 (Three-way formal fuzzy concepts [1–3, 24]). Let us suppose a set of attributes, that is, $(B) = \{y_j, (T_B(y_j), I_B(y_j), F_N(y_j)) \in [0,1]^3 : \forall y_j \in Y\}$ where $j \leq m$. For the selected three-polar attribute set, find their covering object set in the given fuzzy context as:

$$(A) = \left\{x_i, (T_A(x_i), I_A(x_i), F_A(x_i)) \in [0,1]^3 : \forall x_i \in X\right\} \text{ where } i \leq n.$$

The obtained pair (A, B) is called a three-way fuzzy concept iff $A^{\uparrow} = B$ and $B^{\downarrow} = A$. It can be interpreted as a neutrosophic set of objects having maximal truth membership value and minimum indeterminacy and minimum falsity membership values with respect to integrating the information from the common set of fuzzy attributes in a defined three-way space $[0, 1]^3$ using component-wise Gödel-residuated lattice. After that, none of the fuzzy set of objects (or attributes) can be found that can make the membership value of the obtained fuzzy set of attributes (or objects) larger. The pair of neutrosophic set (A, B) is called as formal concepts, where A is extent and B is intent. This formal concept can be visualized as a node of the three-way complete lattice. In this process, a major problem arises when the fluctuation in the three-way concepts occurs at each interval of time. To deal with this problem, we propose a method in the next section using a three-way complex neutrosophic set and its connection with applied abstract algebra [55].

3 Proposed section

In the last year, many researchers have analyzed data with three-way fuzzy attributes and addressed that their periodic measurement at given phases of time is a crucial task for knowledge discovery and representation tasks. To solve this issue, some researchers tried to characterize the uncertainty in fuzzy attributes based on true, indeterminacy, and falsity membership values of a defined neutrosophic set [1–3, 24] and its graphs [16, 52]. In this process, a major problem was addressed while representing the uncertainty and vagueness in attribute changes at each phase of time. To deal with this problem, the chapter introduces a method for precise representation of uncertainty and its fluctuation at a given phase of time using the amplitude and phase term of a complex neutrosophic set. In addition, three methods are proposed to extract some meaningful information from the given three-way complex neutrosophic context based on the user requirements to solve particular problems. The first method focuses on investigating all the complex fuzzy concepts based on their next neighbors, whereas the second one focuses on finding some of the δ-equal complex fuzzy concepts. The third one focuses on decomposition of three-way complex fuzzy contexts at different granulation for rigorous analysis.

3.1 A proposed method for generating three-way complex fuzzy concepts

In this section, a method is proposed for discovery of three-way complex fuzzy concepts based on their next neighbors as given below:

Step 1 The first complex neutrosophic concepts can be investigated by exploring all the objects in the set $\uparrow$

$$\left(x_i, \left(r_{R_{x_i}}e^{w^r_{T_{x_i}}}, r_{I_{x_i}}e^{w^r_{I_{x_i}}}, r_{F_{x_i}}e^{w^r_{F_{x_i}}}\right)\right)^{\uparrow} = \left(y_j, \left(r_{R_{y_j}}e^{w^r_{T_{y_j}}}, r_{I_{y_j}}e^{w^r_{I_{y_j}}}, r_{F_{y_j}}e^{w^r_{F_{y_j}}}\right)\right).$$

The membership value for the complex neutrosophic set of attributes can be computed as follows:

Amplitude
  min $(y_j, |r_{T_{y_j}}|)$ for true membership,
  max $(y_j, |r_{I_{y_j}}|)$ for indeterminacy membership, and
  max $(y_j, |r_{F_{y_j}}|)$ for false membership.

Phase term
  min $(y_j, |e^{w^r_{T_{x_j}}}|)$ for true phase term,
  max $(y_j, |e^{w^r_{I_{y_j}}}|)$ for indeterminate phase term, and
  max $(y_j, |e^{w^r_{F_{y_j}}}|)$ for false phase term.

Similarly, apply the down arrow and find its covering objects to investigate the first concept.

Step 2 The lower neighbor of the complex fuzzy concepts generated at Step 1 can be investigated using uncovered attributes: $y_k = Y - y_j$ where $j \leq m$ and $k \leq m|$.

Step 3 The obtained complex neutrosophic set of attributes can be explored using the Galois connection ($\downarrow$) on amplitude $= (1.0, 0.0, 0.0)$ and phase $= (0, 2\pi)$ term. The covering objects set can be found by ($\downarrow$) as follows:

$$\left(y_j, \left(r_{R_{y_j}}e^{w^r_{T_{y_j}}}, r_{I_{y_j}}e^{w^r_{I_{y_j}}}, r_{F_{y_j}}e^{w^r_{F_{y_j}}}\right)\right)^{\downarrow} = \left(x_i, \left(r_{R_{x_i}}e^{w^r_{T_{x_i}}}, r_{I_{x_i}}e^{w^r_{xx_i}}, r_{F_{x_i}}e^{w^r_{F_{x_i}}}\right)\right).$$

Compute the membership values for the obtained objects:

Amplitude:
  min$(x_i, |r_{T_{x_i}}|)$ for true membership,
  max$(x_i, |r_{I_{x_i}}|)$ for Indeterminacy membership, and
  max$(x_i, |r_{F_{x_i}}|)$ for false membership.

Phase term:
  min$(x_i, |e^{w^r_{T_{x_i}}}|)$ for true phase term,
  max$(x_i, |e^{w^r_{I_{x_i}}}|)$ for indeterminate phase term, and
  max$(x_i, |e^{w^r_{F_{x_i}}}|)$ for false phase term.

Step 4 Apply the up operator ↑ on the constituted objects set:

$$\left(x_i, \left(r_{R_{x_i}}e^{w^r_{T_{x_i}}}, r_{I_{x_i}}e^{w^r_{I_{x_i}}}, r_{F_{x_i}}e^{w^r_{F_{x_i}}}\right)\right)^{\uparrow} = \left(y_j, \left(r_{R_{y_j}}e^{w^r_{T_{y_j}}}, r_{I_{y_j}}e^{w^r_{I_{y_j}}}, r_{F_{y_j}}e^{w^r_{F_{y_j}}}\right)\right).$$

Compute the neutrosophic membership value for the obtained attributes as shown in Step 1.

Step 5 The obtained pair of the complex neutrosophic set of objects and attributes (A, B) represents the lower neighbor of the given concept. The distinct lower neighbors having maximal acceptance of complex neutrosophic membership value while integrating the information among objects and attributes set can be considered as next neighbors.

Step 6 Similarly, all the complex neutrosophic concepts can be discovered using the uncovered attributes.

Step 7 The complex neutrosophic concepts lattice can be built using their next neighbors.

Step 8 Extract some of the meaningful information from the obtained lattice. The pseudo code for the proposed algorithm is shown in Table 7.

Table 7 A proposed algorithm for generating three-way complex neutrosophic concepts.

Input: A three-way complex fuzzy context $K = (X, Y, \widetilde{R})$
where $|X| = n$, $|Y| = m$.
Output: The set of three-way complex fuzzy concepts
1. Find the maximal covering attributes for the object set (X) using $(\uparrow)$:

(i) $\left(x_i, \left(r_{R_{x_i}}e^{w^r_{T_{x_i}}}, r_{I_{x_i}}e^{w^r_{I_{x_i}}}, r_{F_{x_i}}e^{w^r_{F_{x_i}}}\right)\right)^{\uparrow} = \left(y_j, \left(r_{R_{y_j}}e^{w^r_{T_{y_j}}}, r_{I_{y_j}}e^{w^r_{I_{y_j}}}, r_{F_{y_j}}e^{w^r_{F_{y_j}}}\right)\right).$

(ii) Compute the neutrosophic membership value for the obtained attributes:

$\min(y_j, |r_{T_{y_j}}e^{w^r_{T_{x_j}}}|)$ **for true membership,**

$\max(y_j, |r_{I_{y_j}}e^{w^r_{I_{x_j}}}|)$ **for indeterminacy membership, and**

$\max(y_j, |r_{F_{y_j}}e^{w^r_{F_{x_j}}}|)$ **for falsity membership,**

(iii) Apply the operator $(\downarrow)$ on the obtained attribute set:

$\left(y_j, \left(r_{R_{y_j}}e^{w^r_{T_{y_j}}}, r_{I_{y_j}}e^{w^r_{I_{y_j}}}, r_{F_{y_j}}e^{w^r_{F_{y_j}}}\right)\right)^{\downarrow} = \left(x_i, \left(r_{R_{x_i}}e^{w^r_{T_{x_i}}}, r_{I_{x_i}}e^{w^r_{I_{x_i}}}, r_{F_{x_i}}e^{w^r_{F_{x_i}}}\right)\right)$

(iv) This gives first complex neutrosophic concept (A, B).
2. Find its lower neighbor:
3. for $(k = 0$ to $m)$

$Y_k = Y - y_j$ where $j, k \leq m$

(i) New attribute set: $y_k = \{y_j, y_k\}$
(ii) Set maximal acceptance for the complex neutrosophic attributes, i.e., Amplitude = (1.0, 0.0, 0.0) and Phase = $(0, 2\pi)$
(iii) Apply the operator ($\downarrow$) on the attributes

$$\left(y_j, \left(r_{R_{y_j}} e^{w^r_{T_{y_j}}}, r_{I_{y_j}} e^{w^r_{I_{y_j}}}, r_{F_{y_j}} e^{w^r_{F_{y_j}}}\right)\right)^{\downarrow} = \left(x_i, \left(r_{R_{x_i}} e^{w^r_{T_{x_i}}}, r_{I_{x_i}} e^{w^r_{x_{x_i}}}, r_{F_{x_i}} e^{w^r_{F_{x_i}}}\right)\right)$$

(iv) Compute the membership of the obtained objects using Step 1 (ii):
(v) Apply the operator ($\uparrow$) on the constituted set of objects:

$$\left(x_i, \left(r_{R_{x_i}} e^{w^r_{T_{x_i}}}, r_{I_{x_i}} e^{w^r_{x_{x_i}}}, r_{F_{x_i}} e^{w^r_{F_{x_i}}}\right)\right)^{\uparrow} = \left(y_j, \left(r_{R_{y_j}} e^{w^r_{T_{y_j}}}, r_{I_{y_j}} e^{w^r_{I_{y_j}}}, r_{F_{y_j}} e^{w^r_{F_{y_j}}}\right)\right).$$

(vi). Compute the membership of obtained attributes as per Step 1 (ii).
End **for**.
4. Distinct lower neighbor is considered as next neighbor.
5. Similarly, generate all the next neighbor using uncovered attributes.
6. Build the complex neutrosophic concept lattice for knowledge extraction.

Complexity: Let us suppose that the number of objects and the number of attributes in the given three-way complex fuzzy context are n and m, respectively. To discover the lower neighbor of three-way complex fuzzy attributes takes $O(n^3 m)$ time complexity for the amplitude and phase term, respectively. The removal of similar lower neighbor takes at most $O(n^3 * m^3)$ time complexity for the amplitude and phase term, independently. This computation gives the proposed method takes $O(|C|n^6 m^6)$ where, C is lower neighbor. In this way, the proposed method shown in Table 7 takes less computation when compared to any of the available approaches for processing the three-way complex fuzzy dataset.

3.2 A method for finding δ-equal three-way complex fuzzy concepts

It can be observed that the proposed method shown in Table 7 may provide repeated three-way complex fuzzy concepts, which may create a problem in knowledge processing tasks. In this case, choosing some of the interesting complex fuzzy patterns based on the user-required information granules is a computationally expensive task. To achieve this goal, a method is proposed in this section for finding δ-equal complex fuzzy concepts using the distance metric [36, 47]. The steps of the proposed method are given as follows:

Step 1 Let us suppose that a user or expert wants to select a three-way complex fuzzy concept $C_1 = (A_1, B_1)$ having similar amplitude and phase term for the knowledge processing tasks.

Step 2 To investigate this concept the expert can use any of the three-way complex fuzzy concepts from the given complex neutrosophic context, $C_2 = (A_2, B_2)$.

Step 3 The distance among these two complex neutrosophic concepts can be computed to find their semantic similarity. To fulfill this need, the expert can use either the extent or intent of the given concepts as both reflect uniqueness of concepts. The proposed method considers intent of the concepts $(r_{T_{B_1}} e^{i \arg T_{B_1}}, r_{I_{B_1}} e^{i \arg I_{B_1}}, r_{F_{B_1}} e^{i \arg F_{B_1}})$, and $(r_{T_{B_2}} e^{i \arg T_{B_2}}, r_{I_{B_2}} e^{i \arg I_{B_2}}, r_{F_{B_2}} e^{i \arg F_{B_2}})$ as given below:

$$d(B_1, B_2) = \max(\sup(|r_{T_{B_1}} - r_{T_{B_2}}|, |r_{I_{B_1}} - r_{I_{B_2}}|, |r_{F_{B_1}} - r_{F_{B_2}}|)),$$
$$\frac{1}{2\pi} \sup(|\arg_{T_{B_1}} - \arg_{T_{B_2}}|, |\arg_{I_{B_1}} - \arg_{I_{B_2}}|, \sup(|\arg_{F_{B_1}} - \arg_{F_{B_2}}|)).$$

Step 4 The total distance can be computed using the sum of all the distances among attributes, $\sum d(B_1, B_2)$.

Step 5 The average distance among the given concepts can be computed as follows:
$AV(d) = \frac{\sum d(B_1, B_2)}{m}$ where m is total number of attributes in the given concepts.

Step 6 Define a δ-granulation for deciding the similarity level among the three-way complex fuzzy concepts.

Step 7 The complex fuzzy concept C_2 can be chosen iff: $d(C_1, C_2) \leq 1 - \delta$.

Step 8 Similarly, other δ-equal concepts can be investigated.

Complexity: The proposed algorithm shown in Table 8 provides a way to investigate the δ-equal three-way complex neutrosophic concepts. To fulfill this objective the proposed method computes the distance among amplitude and phase term of the given attributes of the concepts, which may take m^3 computational time, independently. In this way the overall time complexity of the proposed method cannot exceed $O(m^6)$. This reduced complexity helps the expert in quick analysis of concept learning when compared to any other approaches.

3.3 A method for decomposition of three-way complex fuzzy context

The previous two methods shown in Tables 7 and 8 provide two alternative ways for processing the given three-way complex fuzzy concepts. In this process, a problem arises when the user wants to navigate or decompose the given three-way complex fuzzy context at different granulation. To achieve this goal, we

Table 8 A proposed algorithm for δ-equal three-way complex fuzzy concepts.

Input: A complex fuzzy concept $(C_1 = (A_1, B_1))$ and $(C_2 = (A_2, B_2))$.
Output: δ-equal three-way complex fuzzy concepts.
1. Enter the given complex fuzzy concepts:
$(C_1 = (A_1, B_1))$, i.e., $r_{B_2}(y) \cdot e^{j \arg_{B_2}(y)}$.
2. Define information granules:
$(C_2 = (A_2, B_2))$, i.e., $r_{B_2}(y) \cdot e^{j \arg_{B_2}(y)}$.
3. Compute the distance among them as follows $d(B_1, B_2)$:
True membership function:
$\max(\sup(|r^T_{B_1}(y) - r^T_{B_2}(y)|, \frac{1}{2\pi}\sup|\arg^T_{B_1}(y) - \arg^T_{B_2}(y)|))$
Indeterminacy membership values:
$\max(\sup(|r^I_{B_1}(y) - r^I_{B_2}(y)|, \frac{1}{2\pi}\sup|\arg^I_{B_1}(y) - \arg^I_{B_2}(y)|))$
Falsity membership values:
$\max(\sup(|r^F_{B_1}(y) - r^F_{B_2}(y)|, \frac{1}{2\pi}\sup|\arg^F_{B_1}(y) - \arg^F_{B_2}(y)|))$
4. Compute the summation of each distance, i.e., $\sum d(B_1, B_2)$.
5. Find the average distance $AV(d) = \frac{\sum d(B_1, B_2)}{m}$.
6. Choose a δ-granulation.
7. if $(d(C_1, C_2) \leq 1 - \delta)$
    Extract the given complex fuzzy concepts.
   else
    Do not select the concept.
8. Similarly, extract all the δ-equal complex fuzzy concepts.

tried to utilize the properties of granular computing. The granulation is an umbrella term that includes many different mathematical ways to process large or complex data via a small chunk of information [8, 9, 38]. This small chunk of information provides a simpler solution for the given problem with improved descriptions. It means granular computing provides a numeric membership function of given complex neutrosophic sets to offer a synthetic and qualitative analysis of context to ensure the process at micro and macro levels. Due to this, its calculus recently applied for decomposition of three-way fuzzy context [2] at distinct multigranulation [3, 37, 39]. This chapter focuses on extending that method for decomposition of the three-way complex fuzzy matrix as shown in Table 9.

Table 9 A proposed algorithm for decomposition of three-way complex fuzzy context.

Input: A three-way complex fuzzy context $\mathbf{K} = (X, Y, \tilde{R})$.

Output: A decomposed context for the chosen complex granules.
1. Write the three-way complex fuzzy context $\mathbf{K} = (X, Y, \tilde{R})$.
2. Define the complex granules (CG) as follows:
$$CG = (r_{T_{\tilde{R}}}(x,y) \cdot e^{i \arg T_{\tilde{R}}(x,y)}, r_{I_{\tilde{R}}}(x,y) \cdot e^{i \arg N_{\tilde{R}}(x,y)}, r_{F_{\tilde{R}}}(x,y) \cdot e^{i \arg N_{\tilde{R}}(x,y)}).$$
3. if

True membership values
$$r_{T_{\tilde{R}}}(x,y) \cdot e^{i \arg T_{\tilde{R}}(x,y)} \geq r_{T_{\tilde{R}}}(x_i, y_j) \cdot e^{i \arg T_{\tilde{R}}(x_i, y_j)}.$$
represent that entries as 1 in the complex neutrosophic matrix.
4. **else** represent the entries as 0.
5. if

Indeterminate membership values
$$r_{I_{\tilde{R}}}(x,y) \cdot e^{i \arg I_{\tilde{R}}(x,y)} \leq r_{I_{\tilde{R}}}(x_i, y_j) \cdot e^{i \arg I_{\tilde{R}}(x_i, y_j)}.$$
represent that entries as 0 in the complex neutrosophic matrix.
6. **else** represent the entries as 1.
7. if

Falsity membership values
$$r_{F_{\tilde{R}}}(x,y) \cdot e^{i \arg F_{\tilde{R}}(x,y)} \leq r_{F_{\tilde{R}}}(x_i, y_j) \cdot e^{i \arg F_{\tilde{R}}(x_i, y_j)}.$$
represent that entries as 0 in the complex neutrosophic matrix.
8. **else** represent the entries as 1.
9. Write the decomposed three-way context.
10. The obtained matrix satisfies following properties:
(i) In case $\delta_1 \leq \delta_2$:
$K_{\delta_2} \subseteq K_{\delta_1}$.
(ii) $K_{\delta_1} \cup K_{\delta_2} \cup \cdots K_{\delta_n} = K$.
11. Interpret the obtained context.

Complexity: The proposed algorithm shown in Table 9 provides a way to find several binary contexts based on user-defined three-way complex fuzzy granules. To achieve this goal, the proposed method finds the entries in the complex fuzzy matrix that have maximal amplitude and phase term from the chosen complex granules. In this case, the proposed method takes total $m \times n$ searches to find the chosen complex information granules. Hence, the total complexity taken by the proposed method cannot exceed the $O(nm^3)$ or $O(mn^3)$ computational time in case of n—number of objects and m—number of three-way complex fuzzy attributes. In this way, the proposed method reduces the time to process the complex fuzzy context when compared to the proposed method shown in Tables 7 and 8. To validate the extracted information, the knowledge discovered from the proposed method is compared with the other two proposals in this chapter.

4 Illustration

This section illustrates the proposed methods as an example with their comparative analysis to validate the obtained results.

4.1 Three-way complex neutrosophic concept lattice

The precise representation of uncertainty and its fluctuation at a given phase of time in three-way fuzzy attributes is addressed as one of the crucial tasks for data analytics researchers [16, 36, 52, 56]. To deal with this issue, the three-way fuzzy concept lattice [1] and its interval-valued [2, 10, 37] concept lattice was studied for multi-decision [3] analysis based on truth, falsity, and indeterminacy membership values, independently. However, measuring the fluctuation in uncertainty and vagueness at a given phase of time is still unsolved in the three-way fuzzy decision space. To accomplish this task some researchers tried to utilize the calculus of complex fuzzy logic [12, 13] and its mathematical properties [21, 40, 41, 52] with neutrosophic sets [36, 52]. In this regard, a hybrid set called a complex neutrosophic set [36] and its graphical structure [52] is introduced for further processing. This chapter puts forward an effort to analyze the three-way complex fuzzy attributes and their hierarchical order visualization in the concept lattice. To fulfill this objective, a method is proposed in Table 7 for generating all the three-way complex fuzzy concepts and their line diagram using the properties of next neighbor. To demonstrate the proposed algorithm, we provide an example in the following section.

Example 6. Let us suppose that an expert wants to analyze the fluctuation in acceptation, rejection, and uncertainty levels of water and its parameters in given areas (x_1, x_2, x_3, x_4). To do this, the expert can collect level of pH (y_1), algae (y_2), conductivity (y_3), dissolved oxygen (y_4), phytoplankton (y_4), turbidity (y_6), precipitation (y_7), alkalinity (y_8), carbon dioxide (y_9), and air and water temperature ($y_1 0$). The expert may select any of the potential parameters to measure the fluctuation in water quality in a particular area. In this chapter, we select the first three parameters for explanation. In case the expert observed that the pH level is 50% accepted in the area x_1 for drinking at 9–10 months, 30% unaccepted after 18 months, uncertain about at 30% after 1 year and

5 months. This three-way complex linguistics word can be written precisely using the properties of a complex neutrosophic set as follows: $x_1 = (0.5e^{i0.7\pi}, 0.3e^{i1.6\pi}, 0.3e^{i1.4\pi})/y_1$.

Similarly, the opinion of the expert for the acceptance of pH (y_1), algae (y_2), and conductivity (y_3) in the given area x_1, x_2, x_3, x_4, as shown in Table 10 whereas Table 11 represents indeterminacy and Table 12 represents its rejection for drinking the water. Table 13 shows the composed representation of this three-way complex fuzzy context using the properties of a complex neutrosophic set. Now the problem is to find some interesting patterns to know the water quality based on chosen parameters.

Step 1 The proposed algorithm shown in Section 3.1 starts the investigation for three-way complex fuzzy concepts using those attributes that cover the objects set maximally. The attribute that covers the object set maximally, $\{(1.0, 1.0)/x_1 + (1.0, 1.0)/x_2 + (1.0, 1.0)/x_3 + (1.0, 1.0)/x_4\}$, can be found using the operator ($\uparrow$) as

Table 10 A true complex membership value of expert opinion about water quality.

	y_1	y_2	y_3
x_1	$0.5e^{i0.7\pi}$	$0.8e^{i1.7\pi}$	$0.4e^{i0.4\pi}$
x_2	$0.3e^{i0.5\pi}$	$0.4e^{i0.3\pi}$	$0.5e^{i0.4\pi}$
x_3	$0.4e^{i1.5\pi}$	$0.6e^{i1.6\pi}$	$0.3e^{i0.5\pi}$
x_4	$0.4e^{i0.2\pi}$	$0.2e^{i0.9\pi}$	$0.7e^{i1.2\pi}$

Table 11 An indeterminacy complex membership value of expert opinion about water quality.

	y_1	y_2	y_3
x_1	$0.3e^{i1.6\pi}$	$0.7e^{i1.1\pi}$	$0.5e^{i0.2\pi}$
x_2	$0.5e^{i1.3\pi}$	$0.1e^{i0.8\pi}$	$0.4e^{i1.4\pi}$
x_3	$0.6e^{i1.9\pi}$	$0.6e^{i1.2\pi}$	$0.3e^{i1.4\pi}$
x_4	$0.7e^{i0.2\pi}$	$0.1e^{i0.5\pi}$	$0.2e^{i0.7\pi}$

Table 12 A falsity complex membership value of expert opinion about water quality.

	y_1	y_2	y_3
x_1	$0.3e^{i1.4\pi}$	$0.2e^{i0.5\pi}$	$0.4e^{i0.7\pi}$
x_2	$0.4e^{i1.3\pi}$	$0.4e^{i1.7\pi}$	$0.3e^{i0.5\pi}$
x_3	$0.5e^{i0.2\pi}$	$0.8e^{i0.9\pi}$	$0.4e^{i1.5\pi}$
x_4	$0.2e^{i0.5\pi}$	$0.9e^{i1.9\pi}$	$0.4e^{i0.2\pi}$

Table 13 A three-way complex neutrosophic context representation of Tables 10–12.

	y_1	y_2	y_3
x_1	$(0.5e^{i0.7\pi}, 0.3e^{i1.6\pi}, 0.3e^{i1.4\pi})$	$(0.8e^{i1.7\pi}, 0.7e^{i1.1\pi}, 0.2e^{i0.5\pi})$	$(0.4e^{i0.4\pi}, 0.5e^{i0.2\pi}, 0.4e^{i0.7\pi})$
x_2	$(0.3e^{i0.5\pi}, 0.5e^{i0.4\pi}, 0.4e^{i1.3\pi})$	$(0.4e^{i0.3\pi}, 0.1e^{i0.8\pi}, 0.4e^{i1.7\pi})$	$(0.5e^{i0.4\pi}, 0.4e^{i1.4\pi}, 0.3e^{i0.5\pi})$
x_3	$(0.4e^{i1.5\pi}, 0.6e^{i1.9\pi}, 0.5e^{i0.2\pi})$	$(0.6e^{i1.6\pi}, 0.6e^{i1.2\pi}, 0.8e^{i0.9\pi})$	$(0.3e^{i0.5\pi}, 0.3e^{i1.4\pi}, 0.4e^{i1.5\pi})$
x_4	$(0.4e^{i0.2\pi}, 0.7e^{i0.2\pi}, 0.2e^{i0.5\pi})$	$(0.2e^{i0.9\pi}, 0.1e^{i0.5\pi}, 0.9e^{i1.9\pi})$	$(0.7e^{i1.2\pi}, 0.2e^{i0.7\pi}, 0.4e^{i0.2\pi})$

$$\{(1.0e^{i2\pi}, 0.0e^{i2\pi}, 0.0e^{i2\pi})/x_1 + (1.0e^{i2\pi}, 0.0e^{i2\pi}, 0.0e^{i2\pi})/x_2$$
$$+ (1.0e^{i2\pi}, 0.0e^{i2\pi}, 0.0e^{i2\pi})/x_3 + (1.0e^{i2\pi}, 0.0e^{i2\pi}, 0.0e^{i2\pi})/x_4\}^{\uparrow}$$
$$= \{(0.3e^{i0.4\pi}, 0.7e^{i1.9\pi}, 0.5e^{i1.3\pi})/y_1 + (0.2e^{i0.3\pi}, 0.7e^{i1.2\pi}, 0.7e^{i1.9\pi})/y_2$$
$$+ (0.3e^{i0.4\pi}, 0.5e^{i1.4\pi}, 0.4e^{i1.5\pi})/y_3\}.$$

Now, apply the operator $\downarrow$ to find the maximal vague set of objects while integrating the information from these attributes as

$$\{(0.3e^{i0.4\pi}, 0.7e^{i1.9\pi}, 0.5e^{i1.3\pi})/y_1 + (0.2e^{i0.3\pi}, 0.7e^{i1.2\pi}, 0.7e^{i1.9\pi})/y_2$$
$$+ (0.3e^{i0.4\pi}, 0.5e^{i1.4\pi}, 0.4e^{i1.5\pi})/y_3\}^{\downarrow} = \{(1.0e^{i2\pi}, 0.0e^{i2\pi}, 0.0e^{i2\pi})/x_1$$
$$+ (1.0e^{i2\pi}, 0.0e^{i2\pi}, 0.0e^{i2\pi})/x_2 + (1.0e^{i2\pi}, 0.0e^{i2\pi}, 0.0e^{i2\pi})/x_3$$
$$+ (1.0e^{i2\pi}, 0.0e^{i2\pi}, 0.0e^{i2\pi})/x_4\}.$$

This provides the following three-way complex neutrosophic concepts:

 (1) Extent:

$$\{(1.0e^{i2\pi}, 0.0e^{i2\pi}, 0.0e^{i2\pi})/x_1 + (1.0e^{i2\pi}, 0.0e^{i2\pi}, 0.0e^{i2\pi})/x_2$$
$$+ (1.0e^{i2\pi}, 0.0e^{i2\pi}, 0.0e^{i2\pi})/x_3 + (1.0e^{i2\pi}, 0.0e^{i2\pi}, 0.0e^{i2\pi})/x_4\}.$$

Intent:

$$\{(0.3e^{i0.4\pi}, 0.7e^{i1.9\pi}, 0.5e^{i1.3\pi})/y_1 + (0.2e^{i0.3\pi}, 0.7e^{i1.2\pi}, 0.7e^{i1.9\pi})/y_2$$
$$+ (0.3e^{i0.4\pi}, 0.5e^{i1.4\pi}, 0.4e^{i1.5\pi})/y_3\}.$$

Step 2 The lower neighbors of concepts shown in Step 1 can be found as follows:

(i)

$$\{(0.3e^{i0.4\pi}, 0.7e^{i1.9\pi}, 0.5e^{i1.3\pi})/y_1 + (0.2e^{i0.3\pi}, 0.7e^{i1.2\pi}, 0.7e^{i1.9\pi})/y_2$$
$$+ (0.3e^{i0.4\pi}, 0.5e^{i1.4\pi}, 0.4e^{i1.5\pi})/y_3\} \cup \{(1.0e^{i2\pi}, 0.0e^{i2\pi}, 0.0e^{i2\pi})/y_1\}.$$

It provides $\{(1.0e^{i2\pi}, 0.0e^{i2\pi}, 0.0e^{i2\pi})/y_1 + (0.2e^{i0.3\pi}, 0.7e^{i1.2\pi}, 0.7e^{i1.9\pi})/y_2 + (0.3e^{i0.4\pi}, 0.5e^{i1.4\pi}, 0.4e^{i1.5\pi})/y_3\}$.

(ii)

$$\{(0.3e^{i0.4\pi}, 0.7e^{i1.9\pi}, 0.5e^{i1.3\pi})/y_1 + (0.2e^{i0.3\pi}, 0.7e^{i1.2\pi}, 0.7e^{i1.9\pi})/y_2$$
$$+ (0.3e^{i0.4\pi}, 0.5e^{i1.4\pi}, 0.4e^{i1.5\pi})/y_3\} \cup \{(1.0e^{i2\pi}, 0.0e^{i2\pi}, 0.0e^{i2\pi})/y_2\}.$$

It provides: $\{(0.3e^{i0.4\pi}, 0.7e^{i1.9\pi}, 0.5e^{i1.3\pi})/y_1 + (1.0e^{i2\pi}, 0.0e^{i2\pi}, 0.0e^{i2\pi})/y_2 + (0.3e^{i0.4\pi}, 0.5e^{i1.4\pi}, 0.4e^{i1.5\pi})/y_3\}$.

(iii)

$$\{(0.3e^{i0.4\pi}, 0.7e^{i1.9\pi}, 0.5e^{i1.3\pi})/y_1 + (0.2e^{i0.3\pi}, 0.7e^{i1.2\pi}, 0.7e^{i1.9\pi})/y_2$$
$$+ (0.3e^{i0.4\pi}, 0.5e^{i1.4\pi}, 0.4e^{i1.5\pi})/y_3\} \cup \{(1.0e^{i2\pi}, 0.0e^{i2\pi}, 0.0e^{i2\pi})/y_3\}.$$

It provides $\{(0.3e^{i0.4\pi}, 0.7e^{i1.9\pi}, 0.5e^{i1.3\pi})/y_1 + (0.2e^{i0.3\pi}, 0.7e^{i1.2\pi}, 0.7e^{i1.9\pi})/y_2 + (1.0e^{i2\pi}, 0.0e^{i2\pi}, 0.0e^{i2\pi})/y_3\}$.

Now the following lower neighbor can be generated from the previously obtained complex neutrosophic attribute using the Galois connection (as illustrated in Step 1):

(2) Extent:

$$\{(0.5e^{i0.7\pi}, 0.3e^{i1.6\pi}, 0.3e^{i1.4\pi})/x_1 + (0.3e^{i0.5\pi}, 0.5e^{i0.4\pi}, 0.4e^{i1.3\pi})/x_2$$
$$+ (0.4e^{i1.5\pi}, 0.6e^{i1.9\pi}, 0.5e^{i0.2\pi})/x_3 + (0.4e^{i0.2\pi}, 0.7e^{i0.2\pi}, 0.2e^{i0.5\pi})/x_4\}.$$

Intent:

$$\{(1.0e^{i2\pi}, 0.0e^{i2\pi}, 0.0e^{i2\pi})/y_1 + (0.2e^{i0.3\pi}, 0.7e^{i1.2\pi}, 0.9e^{i1.4\pi})/y_2$$
$$+ (0.3e^{i0.4\pi}, 0.5e^{i1.4\pi}, 0.4e^{i1.5\pi})/y_3\}.$$

(3) Extent:

$$\{(0.8e^{i1.7\pi}, 0.7e^{i1.1\pi}, 0.2e^{i0.5\pi})/x_1 + (0.4e^{i0.3\pi}, 0.1e^{i0.8\pi}, 0.4e^{i1.7\pi})/x_2$$
$$+ (0.6e^{i1.6\pi}, 0.6e^{i1.2\pi}, 0.8e^{i0.9\pi})/x_3 + (0.2e^{i0.9\pi}, 0.1e^{i0.5\pi}, 0.9e^{i1.9\pi})/x_4\}.$$

Fig. 4 A three-way complex neutrosophic concept lattice built at Step 2.

Intent:

$$\{(0.3e^{i0.4\pi}, 0.7e^{i1.9\pi}, 0.5e^{i1.3\pi})/y_1 + (1.02e^{i2\pi}, 0.0e^{i2\pi}, 0.0e^{i1.9\pi})/y_2$$
$$+ (0.3e^{i0.4\pi}, 0.5e^{i1.4\pi}, 0.4e^{i1.5\pi})/y_3\}.$$

(4) Extent:

$$\{(0.4e^{i0.4\pi}, 0.5e^{i0.2\pi}, 0.4e^{i0.7\pi})/x_1 + (0.5e^{i0.4\pi}, 0.4e^{i1.4\pi}, 0.3e^{i0.5\pi})/x_2$$
$$+ (0.3e^{i0.5\pi}, 0.3e^{i1.4\pi}, 0.4e^{i1.5\pi})/x_3 + (0.7e^{i1.2\pi}, 0.2e^{i0.7\pi}, 0.4e^{i0.2\pi})/x_4\}.$$

Intent:

$$\{(0.3e^{i0.4\pi}, 0.7e^{i1.9\pi}, 0.5e^{i1.3\pi})/y_1 + (0.2e^{i0.3\pi}, 0.7e^{i1.2\pi}, 0.7e^{i1.9\pi})/y_2$$
$$+ (1.0e^{i2\pi}, 0.0e^{i2\pi}, 0.0e^{i2\pi})/y_3\}.$$

It can be observed that each of the obtained lower neighbors are distinct. In this case, each of them can be considered as next neighbor as shown in Fig. 4.

Step 3 Similarly, the following concepts can be generated using the next neighbor of the concept generated at Step 2:

(5) Extent:

$$\{(0.5e^{i0.7\pi}, 0.7e^{i1.6\pi}, 0.3e^{i1.4\pi})/x_1 + (0.3e^{i0.3\pi}, 0.5e^{i0.8\pi}, 0.4e^{i1.7\pi})/x_2$$
$$+ (0.4e^{i1.5\pi}, 0.6e^{i1.9\pi}, 0.8e^{i0.9\pi})/x_3 + (0.2e^{i0.2\pi}, 0.7e^{i0.5\pi}, 0.9e^{i1.4\pi})/x_4\}.$$

Intent:

$$\{(1.0e^{i2\pi}, 0.0e^{i2\pi}, 0.0e^{i2\pi})/y_1 + (1.0e^{i2\pi}, 0.0e^{i2\pi}, 0.0e^{i2\pi})/y_2$$
$$+ (0.3e^{i0.4\pi}, 0.5e^{i1.4\pi}, 0.4e^{i1.5\pi})/y_3\}.$$

(6) Extent:

$$\{(0.4e^{i0.4\pi}, 0.5e^{i1.6\pi}, 0.4e^{i1.4\pi})/x_1 + (0.3e^{i0.4\pi}, 0.5e^{i1.4\pi}, 0.4e^{i1.3\pi})/x_2$$
$$+ (0.3e^{i0.5\pi}, 0.6e^{i1.9\pi}, 0.8e^{i1.5\pi})/x_3 + (0.4e^{i0.2\pi}, 0.7e^{i0.7\pi}, 0.9e^{i1.9\pi})/x_4\}.$$

Intent:

$$\{(1.0e^{i2\pi}, 0.0e^{i2\pi}, 0.0e^{i2\pi})/y_1 + (0.2e^{i0.3\pi}, 0.7e^{i1.2\pi}, 0.7e^{i1.9\pi})/y_2$$
$$+ (1.0e^{i2\pi}, 0.0e^{i2\pi}, 0.0e^{i2\pi})/y_3\}.$$

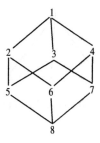

Fig. 5 A three-way complex neutrosophic concept lattice generated from Table 13.

(7) Extent:

$$\{(0.3e^{i0.4\pi}, 0.7e^{i1.1\pi}, 0.5e^{i0.7\pi})/x_1 + (0.4e^{i0.3\pi}, 0.5e^{i1.4\pi}, 0.4e^{i1.7\pi})/x_2$$
$$+ (0.3e^{i0.5\pi}, 0.6e^{i1.2\pi}, 0.8e^{i1.5\pi})/x_3 + (0.2e^{i0.9\pi}, 0.2e^{i0.7\pi}, 0.9e^{i1.4\pi})/x_4\}.$$

Intent:

$$\{(0.3e^{i0.4\pi}, 0.7e^{i1.9\pi}, 0.5e^{i1.3\pi})/y_1 + (1.0e^{i2\pi}, 0.0e^{i2\pi}, 0.0e^{i2\pi})/y_2$$
$$+ (1.0e^{i2\pi}, 0.0e^{i2\pi}, 0.0e^{i2\pi})/y_3\}.$$

(8) Extent:

$$\{(0.4e^{i0.4\pi}, 0.7e^{i1.6\pi}, 0.4e^{i1.4\pi})/x_1 + (0.3e^{i0.3\pi}, 0.5e^{i1.4\pi}, 0.4e^{i1.7\pi})/x_2$$
$$+ (0.3e^{i0.5\pi}, 0.6e^{i1.9\pi}, 0.8e^{i1.5\pi})/x_3 + (0.2e^{i0.2\pi}, 0.7e^{i0.7\pi}, 0.9e^{i1.9\pi})/x_4\}.$$

Intent:

$$\{(1.0e^{i2\pi}, 0.0e^{i2\pi}, 0.0e^{i2\pi})/y_1 + (1.0e^{i2\pi}, 0.0e^{i2\pi}, 0.0e^{i2\pi})/y_2$$
$$+ (1.0e^{i2\pi}, 0.0e^{i2\pi}, 0.0e^{i2\pi})/y_3\}.$$

The above-generated three-way complex fuzzy concepts can be visualized through the vertex and edges of a complex neutrosophic graph as shown in Fig. 5. This line diagram reflects concept 8 as specialized and concept 1 as generalized. The specialized concept 8 represents that the area $(0.4e^{i0.4\pi}, 0.7e^{i1.6\pi}, 0.4e^{i1.4\pi})/x_1$ contains maximal acceptance of water quality for drinking when compared to other areas, whereas the area x_2 stands second. The water in regions x_3 and x_4 is highly unacceptable. Hence, the government needs to change the policy to clean the water of areas x_3 and x_4 first, otherwise it needs to provide some alternative. The generalized concept 1 shows that water in each of the areas is unaccepted for drinking. This analysis will be helpful for the water industry to develop equipment that can control the water quality of a particular area based on given pollutants. At the same time, it will help for government agencies to change policies for citizens of particular areas. In this way, other complex neutrosophic concepts can be analyzed in depth for further processing to solve the particular problem. In this process, a problem may arise when the large number of three-way complex fuzzy concepts can be generated even for small variances in truth, indeterminacy, and falsity membership values. To solve this problem, a method is proposed in Section 3.2 for choosing δ-equal complex fuzzy concepts. In the next section, we illustrate this method using the same example for validating the obtained results.

4.2 δ-Equal three-way complex neutrosophic concepts

In the previous section, it was observed that the size of concept lattice becomes exponential with regard to number of attributes. In this case, extracting some of the potential or interesting patterns in the complex dataset becomes harder. To deal with this problem, researchers focused on controlling the size of the concept lattice [57, 58] using the calculus of granular computing [14, 15, 38, 39] including the three-way neutrosophic contexts [2]. One reason for this is that the properties of granular computing provide a computing paradigm to analyze the given context based on small information of granules [8, 9, 38]. Recently, it was studied in a complex fuzzy set for finding δ-equalities based on several operators given in [36, 40, 41, 52]. Motivated from these recent studies, this chapter introduces a method for finding some of the important or similar concepts at user-defined complex δ-granulation as shown in Table 6. In this section, the proposed method was illustrated on the same context shown in Table 13. This context is feedback from experts about the production of mobile phones in a given economic year. In general, the company focuses on the production of those mobile phones having maximal acceptance of truth membership value and minimal uncertainty and indeterminacy membership values for the given feedback as given by

$$Q_c = \{(1e^{i2\pi}, 0e^{i2\pi}, 0e^{i2\pi})/y_1 + (1e^{i2\pi}, 0e^{i2\pi}, 0e^{i2\pi})/y_2 + (1e^{i2\pi}, 0e^{i2\pi}, 0e^{i2\pi})/y_3\}.$$

The goal is to find δ-equal complex concepts when compared to company required complex query Q_c. Table 14 shows the δ-equal computing steps for the mobile phones x_1. Similarly, for others, mobile phones can be computed using the proposed method as shown in Table 15.

Table 16 shows the δ-equal complex fuzzy concepts selection based on their computed similarity. It can be observed that the

Table 14 δ-Equality of x_1 from the chosen query.

	y_1	y_2	y_3	Max
$\sup|r^T_{B_1}(y) - r^T_{B_2}(y)|$	0.5	0.2	0.6	0.6
$\frac{1}{2\pi}\sup|\arg^T_{B_1}(y) - \arg^T_{B_1}(y)|$	0.65	0.15	0.3	0.65
$\sup|r^I_{B_1}(y) - r^I_{B_2}(y)|$	0.7	0.3	0.5	0.7
$\frac{1}{2\pi}\sup|\arg^I_{B_1}(y) - \arg^I_{B_1}(y)|$	0.2	0.45	0.9	0.9
$\sup|r^F_{B_1}(y) - r^F_{B_2}(y)|$	0.7	0.8	0.6	0.8
$\frac{1}{2\pi}\sup|\arg^F_{B_1}(y) - \arg^F_{B_1}(y)|$	0.3	0.75	0.65	0.75
$AV(d)$				(0.65 + 0.9 + 0.8) = 0.78
δ-equal				0.29

Table 15 δ-Equality for areas based on given query.

Mobiles	Average distance (AV_d)	δ-equal
x_1	0.78	0.22
x_2	0.83	0.17
x_3	0.79	0.21
x_4	0.9	0.1

Table 16 Selection of δ-equal complex fuzzy concepts at user requirements.

δ-equal	Selected complex fuzzy concepts
$\delta \geq 0.9$	⊘
$\delta \geq 0.22$	x_1
$\delta \geq 0.21$	x_1, x_3
$\delta \geq 0.17$	x_1, x_2, x_3
$\delta \geq 0.1$	x_1, x_1, x_3, x_4

x_1 has maximum similarity when compared to other regions where x_3 and x_4 stand last. In this case, the water quality of region x_1 will be somehow more acceptable than other regions, whereas region x_4 is unacceptable. This extracted information from the δ-equal concepts is concordant to three-way complex fuzzy concepts as shown in the last paragraph of Section 4.1. However, a problem arises with this method when an expert wants to navigate the given three-way complex fuzzy context at different granulation for rigorous analysis of hidden patterns. To solve this problem, we proposed a method in Section 3.3, which we demonstrate in the next section for the given context.

4.3 Decomposition of three-way complex fuzzy contexts at different granulations

Granular-based decomposition of given three-way complex fuzzy context beyond $\{\in, \not\subseteq\}$ is considered one of the major issues by researchers [1–3]. The reason is that properties of granular computing provide an umbrella way to analyze the given context in the cost-effective manner [57, 58]. Due to this, its calculus is applied in three-way complex fuzzy concept lattice [36] and other fields [41, 48]. This chapter focuses on navigating the complex neutrosophic context at user-defined complex granules for finding some an interesting patterns. To achieve this goal, user interested three-way complex granules are shown in Table 17. The given complex granules is used to decompose the context shown in Table 13. For illustration of

Table 17 Some of the interesting three-way granules based on user-required preference.

Three-way granulation	User interest	Mathematical measurement	Three-way complex value
Level 1	Maximally interesting	Maximally positive	$(8.0e^{i2\pi}, 0.1e^{i2\pi}, 0.1e^{i2\pi})$
Level 2	Highly interesting	Highly positive	$(0.6e^{i2\pi}, 0.2e^{i2\pi}, 0.2e^{i2\pi})$
Level 3	Very very interesting	Very positive	$(0.5e^{i2\pi}, 0.3e^{i2\pi}, 0.2e^{i2\pi})$
Level 4	Very interesting	Absolute positive	$(0.4e^{i2\pi}, 0.3e^{i2\pi}, 0.3e^{i2\pi})$
Level 5	Interested	Positive	$(0.3e^{i2\pi}, 0.3e^{i2\pi}, 0.3e^{i2\pi})$
Level 6	Not interested	Not positive	$(0.2e^{i2\pi}, 0.4e^{i2\pi}, 0.4e^{i2\pi})$
Level 7	Use less	Negative	$(0.0e^{i2\pi}, 0.5e^{i2\pi}, 0.5e^{i2\pi})$

Table 18 A three-way complex neutrosophic context decomposition at Level 4.

	y_1	y_2	y_3
x_1	(1, 0, 0)	(1, 1, 0)	(1, 1, 1)
x_2	(0, 1, 1)	(1, 0, 1)	(1, 1, 0)
x_3	(1, 1, 1)	(1, 1, 1)	(0, 0, 1)
x_4	(1, 0, 1)	(0, 0, 1)	(1, 0, 1)

the proposed method, granulation level 4, that is, $(0.4e^{i2\pi}, 0.3e^{i2\pi}, 0.3e^{i2\pi})$, is considered. It includes that entries in the three-way complex neutrosophic context with truth membership values more than 0.5 in the given phase of time can be considered as 1. The entries with the indeterminacy and uncertainty membership values less than 0.3 can be considered as 1 as shown in Table 18.

The decomposed context shown in Table 18 shows that the x_1 contains maximal truth membership value, minimal indeterminacy, and minimal falsity when compared to other mobile phones. In this case, the expert concludes that water quality of region x_1 is acceptable when compared to other regions, whereas region x_4 is totally unacceptable. Similarly, other levels of complex granulation shown in Table 17 can be used to extract some of the meaningful information. The major part is that analysis derived from the decomposed context is also similar to its three-way fuzzy concepts and δ-equality, as shown in the last paragraphs of Sections 4.1 and 4.2, respectively. In this way, the proposed method in this chapter provides an umbrella way to solve the given three-way complex fuzzy context for knowledge processing tasks based on user requirements.

5 Discussions

Measuring periodic or recurring fuzzy attributes is a mathematically rigorous task [11]. To deal with this issue, mathematics of fuzzy sets is defined in the complex argand plane with amplitude [0, 1] and [0, 2π] phase term [20, 53]. It gives a way for precise representation of complex linguistics [18, 36], stock market datasets [35, 44], epidemic [59], time series dataset analysis [45] as well as healthcare or medical data analysis [16]. Complex fuzzy sets simulate uncertainty using amplitude terms, whereas the fluctuations use phase terms. In this way, the properties of complex fuzzy sets provide a well-established mathematical model to deal with complex fuzzy context [40, 41]. In this process, a major problem arises when the complex or dynamic datasets contain attribute values beyond the $\{\in, \notin\}$ boundaries [1–3, 16, 36, 52]. One example is analyzing the hue of a color. It is well known that the hue of color is based on the three-way recurring combination of red-green-blue (RGB). Similarly, analyzing the pattern of any sport is based on periodic measurement of three-way spaces (i.e., win, loss, or neutral in the given time interval). In other ways, the given complex fuzzy attributes can be defined based on their acceptation, rejection, and uncertainty at the given phase of time [12, 13, 36, 52, 54]. This will provide a contextual representation of a complex dataset for further analysis using the properties of applied lattice theory. Toward this direction, recently, some researchers have paid attention to analysis of data with three-way fuzzy attributes [32, 39, 42, 43], their concept lattice [1] for partial ignorance [2] in attributes using the properties of neutrosophic set [16]. However, few attentions have been paid toward graphical structure visualization of complex neutrosophic set and its granular decomposition. To deal with this problem, this chapter introduced three methods using the properties of complex neutrosophic sets and their graphical properties as shown in Section 3. The first method focused on generating three-way complex fuzzy concepts and their hierarchical order visualization using the next neighbor algorithm as shown in Section 3.1. The second method focused on finding some of the δ-equal three-way complex fuzzy concepts based on their computed distance as shown in Section 3.2. The third method focused on decomposition of three-way complex fuzzy contexts at user-defined complex granules as shown in Section 3.3. It can be observed that each of the proposed methods is distinct in many aspects as shown in Table 19. However, each of them provides an alternative way to process complex fuzzy contexts. In this case, the selection of a suitable method is based on user requirements to solve the particular problem of a three-way complex fuzzy context and its complexity.

Table 19 shows that each of the proposed methods in this chapter has some advantages and disadvantages in solving the particular problem of three-way complex fuzzy contexts. In this case, the selection of the method can be based on the user or expert requirements to solve the particular problem of given three-way complex datasets. However, the analysis obtained from the proposed methods is concordant with each other. In case the user wants to visualize the given three-way complex fuzzy

Table 19 Comparison of proposed methods on various parameters.

	Proposed method in Section 3.1	Proposed method in Section 3.2	Proposed method in Section 3.3
Three-way decision space	Yes	Yes	Yes
Three-way complex concepts	Yes	Yes	Yes
Three-way complex lattice	Yes	No	Yes
Complex graph	Yes	No	No
Information from Table 13	x_1 is suitable	x_1 is suitable	x_1 is suitable
Methodology	Next neighbor concepts	Distance metric	Granular computing
Time complexity	$O(|C|n^6m^6)$	$O(m^6)$ or $O(n^6)$	$O(nm^3)$ $O(mn^3)$
Advantages	Lower neighbor	δ-equal concepts	Decomposition of complex context
Disadvantages	Repeated concepts	Selecting δ-granules	Level of granulation

datasets in a line diagram for compact analysis, then the proposed method shown in Section 3.1 will be more suitable, whereas the proposed method shown in Section 3.2 will be more suitable in finding some δ-equal complex fuzzy concepts. The proposed method shown in Section 3.3 will be helpful in navigating the three-way complex fuzzy context based on user required complex information granules for approximating the given problem in an umbrella way. These novel outcomes of the proposed methods distinguish them from any of the available approaches in three-way fuzzy concept lattice as shown in Table 20. In this way, the following significant results of the proposed methods in this chapter can be noted:

(1) It introduces a distinct way to process the three-way complex dataset using the properties of applied lattice theory and a complex neutrosophic graph.

(2) It provides a way to find useful patterns from the given three-way complex fuzzy context based on their next neighbors or δ-equality within $O(|C|n^6m^6)$ and $O(m^6)$ complexity, respectively. Here, C represents the lower neighbors, n represents objects, and m represents attributes.

(3) It gives an umbrella way to decompose the given three-way complex fuzzy context at user-defined granulation to

Table 20 Comparison of the proposed method with recent methods.

	Three-way fuzzy set [6, 16, 39]	Three-way interval set [1, 10, 37]	Complex vague set [2, 35, 42, 44]	Proposed methods in this chapter
Domain	Universe of discourse	Universe of discourse	Universe of discourse	Universe of discourse
Codomain	Three-polar single-valued	Three-polar interval-valued	Unit circle	Three-polar circle
True region	[0, 1]	[0, 1]	[0, 1]	[0, 1]
False region	[0, 1]	[0, 1]	[-1, 0]	[0, 1]
Uncertain regions	[0, 1]	[0, 1]	1-true-false	[0, 1]
Amplitude term	Yes	Yes	Yes	Yes
Phase term	No	No	Yes	Yes
Concept lattice	Yes	Yes	Yes	Yes
Pattern	Yes	Yes	Yes	Yes
Granulation	Yes	Yes	No	Yes
δ-Equal	Yes	Yes	No	Yes
Time complexity	$O(2^m n)$	$O(2^m n)$	$O(2^m n^2)$	$O(nm^3)$

The bold values represent that the proposed method is better than other approaches on the given parameters in the row.

characterize the complex dataset by multilevel views to understand the hidden pattern within $O(mn^3)$.

(4) It is helpful to the water industry to develop different types of models for cleaning water based on distinct pollutant parameters.

The following problems are addressed while dealing with complex neutrosophic graphical visualization:

(i) The visualization become more complex in case of more number of complex neutrosophic attributes.
(ii) The processing of complex neutrosophic attributes in case of multidimensional datasets is another concern.
(iii) It needs to explore applicability of complex neutrosophic contexts in various industries.

In the near future, the author will focus on solving the above-mentioned issues with examples.

6 Conclusions

This chapter established a mathematical model to characterize the uncertainty and existing fluctuation of three-way fuzzy attributes using the hybridization of a complex neutrosophic set. In this regard, one of the methods proposed is to discover some of the complex neutrosophic patterns that exist in the given datasets based on their next neighbors and δ-equal closeness within $O(|C|n^6m^6)$ and $O(m^6)$ time complexity, respectively. In addition, another method was proposed for navigating the complex neutrosophic context at user-defined complex granules within $O(mn^3)$ time complexity with an illustrative example. It was also shown that the knowledge discovered from each of the proposed methods is concordant. In this case, the selection of proposed method depends on user or expert requirements to solve the complexity of a particular problem in a given computational time within depth knowledge. In the future, the author will focus on other potential applications of the proposed methods and their extensive properties for knowledge extraction in various fields.

Acknowledgments

The author thanks the anonymous reviewers for their insightful and constructive comments to improve the quality of this chapter.

References

[1] P.K. Singh, Interval-valued neutrosophic graph representation of concept lattice and its $(\alpha\beta\gamma)$-decomposition. Arab. J. Sci. Eng. 43 (2) (2018) 723–740, https://doi.org/10.1007/s13369-017-2718-5.

[2] P.K. Singh, Complex neutrosophic concept lattice and its applications to air quality analysis, Chaos Solit. Fractals 109 (2018) 206–213.

[3] P.K. Singh, Granular based decomposition of complex fuzzy context and its analysis. Prog. Artif. Intell. (2019), https://doi.org/10.1007/s13748-018-00170-y.

[4] J. Qi, L. Wei, Y. Yao, Three-way formal concept analysis, Lect. Notes Comput. Sci. 8818 (2014) 732–741.

[5] J. Qi, T. Qian, L. Wei, The connections between three-way and classical concept lattices, Knowl.-Based Syst. 91 (2016) 143–151.

[6] Y. Yao, Three-way decision: an interpretation of rules in rough set theory, in: P. Wen, Y. Li, L. Polkowski, Y. Yao, S. Tsumoto, G. Wang (Eds.), RSKT 2009, LNCSvol. 5589, 2009, pp. 642–649.

[7] Y. Yao, Three-way decisions with probabilistic rough sets, Inform. Sci. 180 (2010) 341–353.

[8] W. Pedrycz, Shadowed sets: representing and processing fuzzy sets, IEEE Trans. Syst. Man. Cybern. B Cybern. 28 (1998) 103–109.

[9] W. Pedrycz, Granular computing with shadow sets, in: D. Slezak, G.Y. Wang, M. Szezuka, I. Dntsch, Y.Y. Yao (Eds.), RSFDgrc 2005, LNCS (LNAI) 3641, Springer, Heidelberg, 2013, pp. 23–32.

[10] Y. Yao, Interval sets and three-way concept analysis in incomplete contexts, Int. J. Mach. Learn. Cybern. 8 (1) (2017) 3–20.
[11] Y. Yao, S. Wang, X. Deng, Constructing shadowed sets and three-way approximations of fuzzy sets, Inform. Sci. 412–413 (2017) 132–153.
[12] U. Rivieccio, Neutrosophic logics: prospects and problems, Fuzzy Set. Syst. 159 (2007) 1860–1868.
[13] F. Smarandache, A Unifying Field in Logics Neutrosophy: Neutrosophic Probability, Set and Logic, American Research Press, Rehoboth, 1999.
[14] P.K. Singh, Ch.A. Kumar, Bipolar fuzzy graph representation of concept lattice, Inform. Sci. 288 (2014) 437–448.
[15] P.K. Singh, Complex vague set based concept lattice, Chaos Solit. Fractals 96 (2017) 145–153.
[16] P.K. Singh, Three-way fuzzy concept lattice representation using neutrosophic set, Int. J. Mach. Learn. Cybern. 8 (1) (2017) 69–79.
[17] J.A. Goguen, L-fuzzy sets, J. Math. Anal. Appl. 18 (1967) 145–174.
[18] A.B.D. Ulazeez, M. Alkouri, A.R. Salleh, Complex fuzzy soft multisets, the 2014 UKM FST postgraduate colloquium. in: Proceedings of AIP Conference-vol. 1614, 2014, pp. 955–961, https://doi.org/10.1063/1.4895330.
[19] S. Dick, Toward complex fuzzy logic, IEEE Trans. Fuzzy Syst. 13 (3) (2005) 405–414.
[20] D. Ramot, R. Milo, M. Friedman, A. Kandel, Complex fuzzy sets, IEEE Trans. Fuzzy Syst. 10 (2) (2005) 171–186.
[21] Z.Q. Zhao, S.Q. Ma, Complex fuzzy matrix and its convergence problem research, in: B.-Y. Cao (Ed.), Fuzzy Systems & Operations Research and Management, Springer International, Cham, Switzerland, 2016, pp. 157–162.
[22] Ch.A. Kumar, P.K. Singh, Knowledge representation using formal concept analysis: a study on concept generation, in: B.K. Tripathy, D.P. Acharjya (Eds.), Global Trends in Knowledge Representation and Computational Intelligence, IGI Global Publishers, 2014, pp. 306–336.
[23] B. Ganter, R. Wille, Formal Concept Analysis: Mathematical Foundation, Springer-Verlag, Berlin, 1999.
[24] H. Mao, L. Geng-Mei, Interval neutrosophic fuzzy concept lattice representation and interval-similarity measure, J. Intell. Fuzzy Syst. 33 (2) (2017) 957–967.
[25] A. Burusco, R. Fuentes-Gonzalez, The study of the L-fuzzy concept lattice, Math. Soft Comput. 1 (3) (1994) 209–218.
[26] A. Burusco, R. Fuentes-Gonzales, The study on interval-valued contexts, Fuzzy Set. Syst. 121 (3) (2001) 439–452.
[27] R. Shivhare, Ch.A. Kumar, Three-way conceptual approach for cognitive memory functionalities, Int. J. Mach. Learn. Cybern. 8 (1) (2017) 21–34.
[28] S. Dick, R.R. Yager, O. Yazdanbakhsh, On Pythagorean and complex fuzzy set operations, IEEE Trans. Fuzzy Syst. 24 (2016) 1009–1021.
[29] R. Wille, Restructuring lattice theory: an approach based on hierarchies of concepts, in: I. Rival (Ed.), Ordered Sets NATO Advanced Study Institutes Series, vol. 83, 1982, pp. 445–470.
[30] Y. Yao, An outline of a theory of three-way decisions, in: J. Yao, Y. Yang, R. Slowinski, S. Greco, H. Li, S. Mitra, L. Polkowski (Eds.), RSCTC 2012, LNCSvol. 7413, 2013, pp. 1–17.
[31] T.H. Nguyen, A. Kandel, V. Kreinovich, Complex fuzzy sets: towards new foundations. in: The Ninth IEEE International Conference on Fuzzy Systemsvol. 2, 2000, pp. 1045–1048, https://doi.org/10.1109/FUZZY.2000.839195.
[32] B.Q. Hu, Three-way decision spaces based on partially ordered sets and three-way decisions based on hesitant fuzzy sets, Knowl.-Based Syst. 91 (2016) 16–31.

[33] M. Li, J. Wang, Approximate concept construction with three-way decisions and attribute reduction in incomplete contexts, Knowl.-Based Syst. 91 (2016) 165–178.
[34] X. Li, B. Sun, Y. She, Generalized matroids based on three-way decision models. Int. J. Approx. Reason. 90 (2017) 192–207, https://doi.org/10.1016/j.ijar.2017.07.012.
[35] G. Selvachandran, P.K. Maji, I.E. Abed, A.R. Salleh, Complex vague soft sets and its distance measures, J. Intell. Fuzzy Syst. 31 (2016) 55–68.
[36] M. Ali, F. Smarandache, Complex neutrosophic set, Neural Comput. Appl. 28 (7) (2017) 1817–1834.
[37] L. Ma, J.S. Mi, B. Xie, Multi-scaled concept lattices based on neighborhood systems, Int. J. Mach. Learn. Cybern. 8 (1) (2017) 149–157.
[38] P.K. Singh, Ch.A. Kumar, A method for decomposition of fuzzy formal context, Proc. Eng. 38 (2012) 1852–1857.
[39] J.H. Li, C. Huang, J. Qi, Y. Qian, W. Liu, Three-way cognitive concept learning via multi-granularity, Inform. Sci. 378 (2017) 244–263.
[40] O. Yazdanbakhsh, S. Dick, A systematic review of complex fuzzy sets and logic. Fuzzy Set. Syst. (2017), https://doi.org/10.1016/j.fss.2017.01.010.
[41] G. Zhang, T.S. Dillon, K.Y. Cai, J. Ma, J. Lu, Operation properties and δ-equalities of complex fuzzy sets, Int. J. Approx. Reason. 50 (2009) 1227–1249.
[42] C. Huang, J.H. Li, C. Mei, W.Z. Wu, Three-way concept learning based on cognitive operators: an information fusion viewpoint, Int. J. Approx. Reason. 83 (2017) 218–242.
[43] C. Li, F.T. Chan, Knowledge discovery by an intelligent approach using complex fuzzy sets, Lect. Notes Comput. Sci. 7196 (2012) 320–329.
[44] G. Selvachandrana, P.K. Maji, I.E. Abed, A.R. Salleh, Relations between complex vague soft sets, Appl. Soft Comput. 47 (2016) 438–448.
[45] O. Yazdanbakhsh, S. Dick, Time-series forecasting via complex fuzzy logic, in: A. Sadeghian, H. Tahayori (Eds.), Frontiers of Higher Order Fuzzy Sets, Springer, New York, NY, 2015, pp. 147–165.
[46] L.A. Zadeh, Fuzzy sets, Inf. Control 8 (1965) 33–53.
[47] L.A. Zadeh, The concepts of a linguistic and application to approximate reasoning, Inform. Sci. 8 (1975) 199–249.
[48] L.A. Zadeh, A note on Z-numbers, Inform. Sci. 181 (2011) 2923–2932.
[49] L.Q. Dat, N.T. Thong, M. Ali, F. Smarandache, M. Abdel-Basset, H.V. Long, Linguistic approaches to interval complex neutrosophic sets in decision making, IEEE Access 7 (2019) 38902–38917.
[50] V.E. Balas, S.S. Roy, D. Sharma, P. Samui, Handbook of Deep Learning Applications, vol. 136, Springer, 2019.
[51] R. Biswas, A. Vasan, S. Roy, Dilated deep neural network for segmentation of retinal blood vessels in fundus images. Iranian J. Sci. Technol. Trans. Electr. Eng. (2019), https://doi.org/10.1007/s40998-019-00213-7.
[52] S. Broumi, A. Bakali, M. Talea, F. Smarandache, Complex neutrosophic graphs of type 1, in: Proceedings of IEEE International Conference on Innovations in Intelligent Systems and Applications (INISTA)Gdynia Maritime University, Gdynia, Poland, 2017, pp. 432–437.
[53] D. Ramot, M. Friedman, G. Langholz, A. Kandel, Complex fuzzy logic, IEEE Trans. Fuzzy Syst. 11 (4) (2003) 450–461.
[54] S. Broumi, I. Deli, F. Smarandache, N-valued interval neutrosophic sets and their application in medical diagnosis, Crit. Rev. 10 (2015) 46–69.
[55] M. Ward, R.P. Dilworth, Residuated lattices, Trans. Am. Math. Soc. 45 (1939) 335–354.

[56] P. Gajdos, V. Snasel, A new FCA algorithm enabling analyzing of complex and dynamic data sets, Soft. Comput. 18 (4) (2014) 683–694.
[57] Ch.A. Kumar, S. Srinivas, Concept lattice reduction using fuzzy K-means clustering, Expert Syst. Appl. 37 (3) (2010) 2696–2704.
[58] Ch.A. Kumar, S.M. Dias, N.J. Vieira, Knowledge reduction in formal contexts using non-negative matrix factorization, Math. Comput. Simul. 109 (2015) 46–63.
[59] D.E. Tamir, N.D. Rishe, A. Kandel, Complex fuzzy sets and complex fuzzy logic: an overview of theory and applications, in: D.E. Tamir (Ed.), Fifty Years of Fuzzy Logic and Its Applications, Springer International Publishing, Cham, Switzerland, 2015, pp. 661–681.

Index

Note: Page numbers followed by *f* indicate figures, *t* indicate tables, and *b* indicate boxes.

A

Accuracy
  classification model testing, 184
  for decision tree model, 156–157, 157*f*
  definition of, 8–9
  for deep neural network (DNN) method, 157, 158*f*
AdaBoost algorithm, 151–152*b*
AdaBoostClassifier, 151–152, 156–157
Adam. *See* Adaptive moment optimizer/estimator (Adam)
Adaptive moment optimizer/ estimator (Adam), 210
ADC. *See* Analog-to-digital converter (ADC)
AI. *See* Artificial intelligence (AI)
ALARM-NET system, 123
All India Federation of the Deaf, 30
ALS. *See* Amyotrophic lateral sclerosis (ALS)
American Sign Language, 44
Amyotrophic lateral sclerosis (ALS), 132
  mouse model, 139
  mutant carrier, 134
  patients, 132, 134–139
Analog-to-digital converter (ADC), 165
Android application-based model, 193–194
Android Studio software, 48
Anonymized name entity, 88–89
Aortic stenosis (AS), 218–219
App design, 45–48, 46*f*
App development, 47–48

Application integration (semantic interoperability), 16–17
App requirement analysis, 43–45
Apriori, 179–180
Area under curve (AUC), 185–186
Area under the ROC curve (AUC), 116
Arithmetic proficiency test, of HI and NH students, 35–37, 37*f*, 40–41*t*
Artificial intelligence (AI), 59–60, 68, 70, 168
  in transplantation, 60
Artificial neural networks, 186–187
AS. *See* Aortic stenosis (AS)
Association rule, 179–180
ATHENA interoperability framework (AIF), 16–17
AUC. *See* Area under curve (AUC)
Automatic translation system, 86
Autoregressive (AR) modeling method, heart sound recordings, 220–221

B

Bayesian networks, 121
  model, 188
BCI. *See* Brain-computer interface (BCI)
Behavioral method
  driver drowsiness detection, 168–172
    algorithm, for predicting drowsiness, 170
    eye flicker rate, 169
    flow diagram of, 169*f*
    shut eye examination (SEE), 168–169

  yawning, evaluation of, 170
BERT, 75–76, 76–77*f*
  basic base structure, 95
  basic model, 73–80, 88–89, 93–94, 98
  basic pretraining, 75–76
  biomedical and clinical, 76–98
    bioBERT, 78
    clinical BERT, 78–79
    pretrained models, 77–79
    SciBERT, 79
    text mining, fine-tuning for, 80–96
    discussions, 97–98
    pre-training and fine-tuning procedures, 75–76, 76*f*
    vectored sentence, 96
Bert vectored sentence, 96
Binary classification, 227
BioBERT, 78, 87–92, 90–91*t*
Biomedical domain-specific datasets, 78
Biomedical entity normalization issue, 80–84
Biomedical Language Understanding Evaluation (BLUE), 80, 81–82*t*
Biomedical text corpus, 74, 78
Biosensors, 121
Bio-signal approach, 164
BLUE. *See* Biomedical Language Understanding Evaluation (BLUE)
Brain-computer interface (BCI), 164
Breast cancer, classification model, 187
Breathing sensor, 119–120

269

Index

C

Cancer, 58
The Cancer Imaging Archive (TCIA), 105–106
Cardiac auscultation, 216
Cardiac cycle, 219, 221–222
Cardiometabolic health, 62–67, 66t, 69t
Cardiovascular diseases (CVDs), 58, 215
Categorical cross-entropy loss function, 114
CDR. *See* Chemical-Disease Relations (CDR)
CDSS. *See* Clinical decision support system (CDSS)
Centroid frequency, 217–218, 221–222
CHD. *See* Coronary heart disease (CHD)
Chemical-Disease Relations (CDR), 81–84
Chest radiograph, 206
Chronic obstructivepulmonary disease, 58
CIMI. *See* Cyber Integrated Medical Infrastructure (CIMI)
Circuit diagram, heartbeat detection method, 165–168, 166–167f, 173
Classification model, 179–189
   classifier for, 182
   building, 181, 182f
   for disease prediction, 186–189
   driver drowsiness detection, 170–172
     comparative analysis, 172
     convolutional neural network (CNNs), 172
     hidden Markov models (HMMs), 171
     support vector machines (SVMs), 171
   heart sound recordings
     k-nearest neighbor (k-NN) classifier, 222, 225–226, 227–228t
     long short-term memory (LSTM)network classifier, 223–228, 224f, 227t
     one-dimensional convolutional neural network (1-D-CNN) classifier, 223, 223f, 225–228, 227t
     support vector machine (SVM) classifier, 222, 225–226, 227–228t
   issues
     data cleaning, 182
     data transformation and reduction, 183–184
     relevance analysis, 182–183
   testing
     accuracy, 184
     area under curve (AUC), 185–186
     confusion matrix, 185, 185f
     receiver operating characteristic (ROC) curve, 185–186, 186f
     sensitivity, 184
     specificity, 185
Classifier, 181–182, 182f
Clavicle bone segmentation, 213
   computed tomography (CT) segmentation, 205–206, 209f
   U-Net model
     accuracy, computation of, 210
     construction, 207, 208–209f
     data, 206
     dice similarity coefficient (DSC), 210, 210t, 212f
     network architecture, 206–207
     training and optimization, 208–210
Clinical BERT, 78–79, 92t
   embedding paper, 74, 80, 98
   model, 93–94
   prediction results, 93t
Clinical bioBERT training procedure, 94
Clinical decision support system (CDSS), 20, 23, 68, 70
Clinical practice guidelines (CPG), 23
Clinical text mining research, 74–75
Clustering algorithms, 122
Clustering techniques, 62, 67–68
Clusters, characteristics of, 64–65t
CNNs. *See* Convolutional neural networks (CNNs)
Cohort, 60–61, 61t
Compartmental model, 196–198
   infection rate, 196
   latent rate, 196
   recovery rate, 196
   reproduction number, 196
   susceptible-infectiousrecover (SIR), 196–197
Complex fuzzy set, 238
Complex neutrosophic graph, 241–242, 242–243t, 243f
Complex neutrosophic set, 235–238, 251
Computed tomography (CT)
   clavicle bone segmentation, 205–206, 209f
   scan images classification, 105–106, 116
   convolutional neural networks (CNN), 106–107, 108t
   dataset, 109–110
   2-D CNNs, implementation of, 111–114
   error matrix, 115, 115t
   exploratory data analysis, 110–111
   objectives, 106
   related study, 108–109
   results and discussion, 114–116
Conceptual integration (organizational interoperability), 16–17
Conditional random field (CRF), 95
Confusion matrices

classification model testing, 185, 185f
heart sound recordings, 225, 225–226f
CONTAINS temporal relation extraction task, 84–86, 85t
Context awareness concept, 58–59
Contrast material usage, 105–106
Convolutional neural networks (CNNs), 106–107, 108–109t, 147, 205
  driver drowsiness detection, 172
  technique, 149
  training stage of, 114
Convolutional restricted boltzmann machine, 108–109, 108–109t
Coronary heart disease (CHD), 187–188
CPG. See Clinical practice guidelines (CPG)
Credible clinical context-based word embeddings, 79
CRF. See Conditional random field (CRF)
CT. See Computed tomography (CT)
CVDs. See Cardiovascular diseases (CVDs)
Cyber Integrated Medical Infrastructure (CIMI), 187–188

D

Database creation, steps for, 47, 47f
Data cleaning, 182
Data exchange, syntax of, 17–18
Data mining, 179, 187, 193
  technique, 147–149
Data modeling, 70
Data preparation, 120
Data reduction, 183–184
Data science point, 131
Dataset, 6, 109–110, 110f
  samples, 2f
Data source, 183–184

Data transformation, 183–184
Data visualization technique, 145–146, 152–153
2-D CNNs, implementation of, 111–114, 112f
Decision support system (DSS), 190
Decision tree algorithm, 149–151, 150–151b
Decision tree classifier, 193–194
Decision tree model, disease diagnosis using
  accuracy, 156–157, 157f
  AdaBoostClassifier, 151–152, 151–152b
  decision tree algorithm, 149–151, 150–151b
  implemented method, 152–153
Deep learning, 73, 106–107
  method, 228
Deep neural network (DNN) method, disease diagnosis using
  accuracy, 157, 158f
  activation functions, 153
  architecture, 154t
  implemented method, 153–154
  model, 155f
  rectified linear unit (ReLU), 153, 154f
  softmax, 153
De Morgan algebra, 243
DEP. See Dependency parsing (DEP)
Dependency parsing (DEP), 95
DFT. See Discrete Fourier transform (DFT)
Diabetes mellitus (DM), 192
  factors affecting, 195f
  regression models for, 192–194
Dice similarity coefficient (DSC), 208–210
DICOM image name, 109–110
Discrete Fourier transform (DFT), 220
  heart sound recordings, 220

Diseasedeep belief network, 108–109, 108–109t
Disease prediction
  challenges in, 198–199
  classification models for, 186–189
  predictive model, 189–190
  predictive model validation, 190, 190f
  regression model for, 189–198
  steps involved in, 189, 189f
DM. See Diabetes mellitus (DM)
Domain-specific bert model, 74
Domain-specific natural language processing task, 76
Driver drowsiness detection
  behavioral method, 168–172
  algorithm, for predicting drowsiness, 170
  eye flicker rate, 169
  flow diagram of, 169f
  shut eye examination (SEE), 168–169
  yawning, evaluation of, 170
  classifications for, 170–172
  comparative analysis, 172
  convolutional neural network (CNNs), 172
  hidden Markov models (HMMs), 171
  support vector machines (SVMs), 171
  heartbeat detection method, 165–168, 168f
  Arduino UNO, 165, 166f
  circuit diagram, 165–168, 166–167f
  components, 165
  graphical representation, 167–168, 167f
  grove ear clip, 165
Drowsiness, prediction algorithm for, 170
DSC. See Dice similarity coefficient (DSC)
DSS. See Decision support system (DSS)

E

EANN. *See* Evolutionary artificial neural network (EANN)
EAR. *See* Eye angle ratio (EAR)
Ear clip heart sensor, 165
Ear grove sensor, 165
ECG. *See* Electrocardiogram (ECG) sensor
EEG. *See* Electroencephalogram (EEG) sensor
eHealth, 70
Electrocardiogram (ECG) sensor, 119–120, 124, 215
Electrocardiography (ECG), 164
Electroencephalogram (EEG) sensor, 119–120, 215
Electroencephalography (EEG), 164
Electromyogram (EMG) sensor, 119–120, 215
Electromyography (EMG), 164
Electronic health records (EHRs) systems, 57–58
  interoperability issues in
    benefits of semantic interoperability, 21–22
    challenges in semantic interoperability, 22–23
    domains of, 14
    healthcare systems, 18–19, 20*t*
    information in, 14–15
    ISO3 definition, 14–15
    personal health record (PHR), 13–14
    semantic interoperability in healthcare, 19–21, 21*f*
    types of, 15–18, 16*t*
    web technology, 15
  note, 81–84
Electrooculography (EOG), 164
ELM. *See* Extreme learning machine (ELM)
EMBalance Decision Support System (DSS), 188
EMG. *See* Electromyogram (EMG) sensor
Encoder-decoder convolutional neural network (CNN)
  architecture, 4*f*
  average error for rotational and translational values, 10*t*
  comparison between ground-truth values and predicted values, 9*t*
  dataset, 6
  machine learning algorithms, 2–3
  system overview and analysis, 3–5
  training and evaluation, 6–7, 8*f*
Entity normalization architecture, 80–81, 83*f*
Entity normalization system, 80–81
Entity-type classification task, 86–87
Environmental sensors, 121
EOG. *See* Electrooculography (EOG)
Epidemic models
  deterministic, 195
  stochastic, 195–196
EPP. *See* Extended physiological proprioception (EPP)
Error matrix, 115, 115*t*
European Interoperability Framework, 17, 17*f*
Evolutionary artificial neural network (EANN), 188
Exoskeleton, 122
Exploratory data analysis, 110–111, 110–111*f*
Extended physiological proprioception (EPP), 122
Extreme learning machine (ELM), 147–149
Eye angle ratio (EAR), 168–169
Eye flicker rate, 169

F

fALS. *See* familial ALS (fALS)
False negative, 184
familial ALS (fALS), 134
FE. *See* Feature extraction (FE)
Feature extraction (FE), 132
  heart sound recordings
    autoregressive (AR) modeling method, 220–221
    centroid frequency computation, 221–222
    discrete Fourier transform (DFT) method, 220
    modeling, 220–221
Fine-tune BERT model, 75–76
Finite length discrete-time signal sequence, 220–221
fMRI image separation, 108–109, 108–109*t*
fNIRS. *See* Functional near-infrared spectroscopy (fNIRS)
Fog computing, 124
Foundational interoperability, 17–18
Frame extraction, heart sound recordings, 220
Framingham heart study, 66–67
Framingham risk score (FR), 61, 68
Functional interaction-induced latent source detection, 108–109, 108–109*t*
Functional near-infrared spectroscopy (fNIRS), 164
Fuzzy rule-based classification algorithm, 186
Fuzzy rule-based classifier, 186–187
Fuzzy rule-based neural classifier, 187
Fuzzy rule-based system, 187–188
Fuzzy sets, 235–236, 236*t*

G

Gene expression omnibus (GEO), 134
General Language Understanding Evaluation (GLUE), 80
General-purpose language domain, 75–76
Genetic programming algorithm, 188

GEO. *See* Gene expression omnibus (GEO)
Geriatric medical examination (GME), 188
GLUE. *See* General Language Understanding Evaluation (GLUE)
GME. *See* Geriatric medical examination (GME)
Graphical interface, 122
Grove ear clip, 165

H
HAPUs. *See* Hospital-acquired pressure ulcers (HAPUs)
HC disease classification, 108–109, 108–109*t*
Healthcare information and management systems society (HIMSS), 17–18
Healthcare services, machine learning in
 artificial intelligence (AI), 60
 cardiometabolic health, 62–67, 66*t*
 Cohort, 60–61, 61*t*
 context awareness concept, 58–59
 electronic health records (EHRs), 57–58
 liver transplantation, 59
 Model for End-Stage Liver Disease (MELD) score, 61
 National Liver Transplantation Program, 60–61
 principal component analysis (PCA), 62
 silhouette criterion, 63, 63*f*
 vascular age, 62–67, 66*t*
Healthcare systems, interoperability in, 18–19, 20*t*
Health information systems (HISs), 13
Hearing impaired (HI) persons, 30
 All India Federation of the Deaf, 30
 app design, 45–48, 46*f*
 app requirement analysis, 43–45
 brain of, 32–33, 33*f*
 hypothesis, 34
 mobile application, 48–50, 49–53*f*
 normal hearing (NH) persons, 29–30, 35, 48–50
 problem of, 32–34
 statistical analysis of survey outcome, 40–42, 40–41*t*
 survey methodology, 35–39
 vocabulary knowledge acquirement (VKA), 52–54
Heartbeat, 124
Heartbeat detection method
 driver drowsiness detection, 165–168, 168*f*
 Arduino UNO, 165, 166*f*
 circuit diagram, 165–168, 166–167*f*
 components, 165
 graphical representation, 167–168, 167*f*
 grove ear clip, 165
Heart disease, prediction of, 190–191
Heart rate ear clip, 165
Heart rate sensor, 165
Heart sound analysis, 217, 229
Heart sounds, 216–217
 feature extraction, 216
 recordings
  classification, 222–224
  confusion matrices, 225, 225–226*f*
  database, 218–219, 219*t*
  feature extraction, 220–222
  method, 218–224, 218*f*
  performance evaluation, 224
  preprocessing, 219–220
HI. *See* Hearing impaired (HI) persons
Hidden Markov chain, 171
Hidden Markov models (HMMs), 121
 driver drowsiness detection, 171
Hidden semi-markov model, 121, 124–125
High blood pressure, 192
High blood sugar, 192
High-level automated system, 106–107
HIMSS. *See* Healthcare information and management systems society (HIMSS)
HISs. *See* Health information systems (HISs)
Histogram of sloping gradients (HOG), 170
HMMs. *See* Hidden Markov models (HMMs)
HOG. *See* Histogram of sloping gradients (HOG)
Hospital-acquired pressure ulcers (HAPUs), classification model, prediction using, 188
Hounsfield unit (HU), 111
Hour glass encoder-decoder model, 6*f*
HPM. *See* Hybrid prediction model (HPM)
HU. *See* Hounsfield unit (HU)
Human-labeled training sample, 73
Hybrid prediction model (HPM), 193
Hyperbolic tangential function, 107

I
ICU. *See* Intensive care unit (ICU)
IDF. *See* International Diabetes Federation (IDF)
IHD, 191
Imagesthree-dimensional convolutional neural network, 108–109, 108–109*t*
Indian Institute of Psychometry (IIP), 31
Indian Sign Language (ISL), 30–31

Indian Sign Language Recognition (ISLR), 31
Indian Sign Language Research Training Center (ISLRTC), 47
Indian statistical institute, 44–45
Indian Weighted Diabetes Risk Score (IWDRS), 194
Induced pluripotent stem (iPS), 147–149
Influential pretrained model, 77
Infrared (IR) cameras, 170
Intensive care unit (ICU), 92
International Diabetes Federation (IDF), 193
Internet of Things (IoT) health-care system
  data analytics in
    benefits of, 119–120
    electroencephalogram (EEG) sensor, 119–120
    goal of, 119–120
    off-line data analytic in, 121–124
    online data analytic in, 124–125
    real-time data analysis in, 126
IoT devices, 23
iPS. See Induced pluripotent stem (iPS)
IR. See Infrared (IR) cameras
Ischemic heart disease (IHD), prediction of, 191
ISL. See Indian Sign Language (ISL)
ISLR. See Indian Sign Language Recognition (ISLR)
ISLRTC. See Indian Sign Language Research Training Center (ISLRTC)
ISO3, definition of, 14–15
IWDRS. See Indian Weighted Diabetes Risk Score (IWDRS)

J
Java language, 48

K
Kernel density point, 110
Key-enhanced BERT pretrained model, 80
K-means clustering algorithm, 193
K-nearest neighbor (k-NN), 164, 217, 222
  heart sound recordings, 222, 225–226, 227–228t

L
Language proficiency test, of HI and NH students, 35–36, 38–39, 38–39f
Lasso linear regression model, 147–149
Latent rate, 196
LBPs. See Local binary patterns (LBPs)
LDA. See Linear discriminant analysis (LDA)
Leave-one-out cross validation (LOOCV), 133
Levels of Information Systems Interoperability (LISI) Reference Model, 16, 16t
Linear discriminant analysis (LDA), 133
LISI. See Levels of Information Systems Interoperability (LISI) Reference Model
Liver transplantation, 59
Local binary patterns (LBPs), 170
Logistic regression algorithm, 194
Logistic regression model, 191
Log likelihood function, 79
Long clinical notes input, 98
Long short-term memory network classifier (LSTM)
  forget gate, 224
  heart sound recordings, 223–228, 224f, 227t
  input gate, 224
  memory cell, 224
  output gate, 224
LOOCV. See Leave-one-out cross validation (LOOCV)
Loss function, 7, 113
LSTM. See Long short-term memory network classifier (LSTM)

M
Machine learning, 106–107, 145–146, 222
  algorithms, 2–3
  disease diagnosis using dataset and metrics, 155–156
  decision tree model, 149–153, 150–151b, 156–157, 157f
  deep neural network method, 153–154, 157, 158f
  research, 147–149, 150t
  supervised, 180
  unsupervised, 180
Magnetically actuated soft capsule endoscopes (MASCE), 6
Market basket analysis, 179–180
Markov chain process, 195–196
MARS. See Multivariate adaptive regression splines (MARS)
MASCE. See Magnetically actuated soft capsule endoscopes (MASCE)
Medical Document Anonymization (MEDDOCAN), 86–87
MELD. See Model for End-Stage Liver Disease (MELD) score
Memory cell, 224
MIMIC-III clinical datasets, 84–86
MIMIC-III dataset, 92
miRNA expression profile, 134, 137, 140–142
  for ALS patients
    serum miRNA, 134–139
    skeletal muscle mRNA, 134–135
miRTarBase 2017, 141, 142t

Mitral regurgitation (MR), 218–219
Mitral stenosis (MS), 218–219
Mitral valve prolapse (MVP), 218–219
MLP. *See* Multilayer perceptron (MLP)
MobiCare, 123
Mobile ECG system, 122
Model for End-Stage Liver Disease (MELD) score, 60–61, 68–69
MR. *See* Mitral regurgitation (MR)
mRNAs, 133, 141, 143
  ALS patients
    serum miRNA expression profiles for, 134–139
    skeletal muscle mRNA expression profiles for, 134–135
  PD patients
    substantia nigra mRNA expression profiles for, 134, 140–142
MS. *See* Mitral stenosis (MS)
Multilayered bidirectional transformer encoder, 75
Multilayer perceptron (MLP), 106–107, 193
Multiobjective genetic algorithm, 187–188
Multiple linear regression, 191
Multivariate adaptive regression splines (MARS), 147–149
MVP. *See* Mitral valve prolapse (MVP)
MyPHI, 188

N

Naive Bayes (NB) algorithms, 187
Name entity recognition (NER), 78, 80, 86, 88–89, 95
National Center for Biotechnology Information (NCBI), 81
National liver transplantation program, 60–61
Natural language processing (NLP), 75
  biomedical, 80
  research community, 73, 99
  research field, 73
  task, 73–79, 87, 94t, 98–99
NB. *See* Naive Bayes (NB) algorithms
NCBI. *See* National Center for Biotechnology Information (NCBI)
NCCTG. *See* North Central Cancer Treatment Group (NCCTG)
NER. *See* Name entity recognition (NER)
Network hourglass pose network, 4
Neural network classifier, 188
Neural network framework, 113
Neurodegenerative disease, 143
Neutrosophic set, lattice structure of, 243
Next generating sequencing (NGS), 131
NGS. *See* Next generating sequencing (NGS)
NLP. *See* Natural language processing (NLP)
Nonlinear activating process, 113
Nontechnical summaries (NTSs), 86
Normal hearing (NH) persons, 29–30, 35, 48–50
  brain of, 32–33, 32f
  survey methodology, 35–39
North Central Cancer Treatment Group (NCCTG), 187
NTSs. *See* Nontechnical summaries (NTSs)

O

ODHMAD. *See* Online model for daily habitant and anomaly detection (ODHMAD)
Off-line data analytics, in IoT health-care system, 121–124
Omics data set, 134
One-dimensional convolutional neural network classifier (1-D-CNN), 222–223
  heart sound recordings, 223, 223f, 225–228, 227t
Online data analytics, in IoT health-care systems, 124–125
Online model for daily habitant and anomaly detection (ODHMAD), 124
Optimal hyperplane, 222
Organizational interoperability, 15–18
Original BERT pretrained models, 77–78
Outlier detection, 179–180

P

Pair-wise and size-constrained K-means (PSCKmeans), 194
Pareto differential evolution (PDE) algorithm, 188
Parkinson's disease (PD), 132
  patient, 134, 140–142
    confusion matrix of, 141t
    substantia nigra mRNA expression profiles for, 134, 140–142
Participant Intervention Comparison Outcome (PICO), 95
PCA. *See* Principal component analysis (PCA)
PCA-based unsupervised feature extraction (PCAUFE), 132
  linear discriminant analysis, 133
  materials and methods, 132–134
  omics data sets, 134
  results, 134–142
PCG. *See* Phonocardiogram (PCG)

PC loading, 132–134, 137, 140–141
PDA. *See* Personal digital assistant (PDA)
PDE. *See* Pareto differential evolution (PDE) algorithm
Pediatric controlled computed tomography scan, 105–106
Penalized logistic regression (PLR), 188
Personal digital assistant (PDA), 191
Personal health record (PHR), 13–14, 20, 23
PHI. *See* Protected health information (PHI)
Phonocardiogram (PCG), 215–216
Phonocardiography, 216
PICO. *See* Participant Intervention Comparison Outcome (PICO)
PLR. *See* Penalized logistic regression (PLR)
PMC full text article, 78
Predictive analytics, 70
Pretrained BERT model, 76–79, 86–87
 BioBERT, 78
 clinical BERT, 78–79
 language model, 79
 language representation model, 80
 original, 77–78
 sciBERT, 79
Principal component analysis (PCA), 62, 132, 186
Probabilistic prediction model, 194
Proposed AR-based feature extraction method, 229
Protected health information (PHI), 86–87
Prototype exemplar learning classifier (PEL-C), 188

Q
QA. *See* Questing answering (QA)
Questing answering (QA), 78, 88–89

R
RE. *See* Relation extraction (RE)
Receiver operating characteristics curve (ROC), 116
 curve, 185–186, 186f
Rectified linear unit (ReLU), 113, 153
 layers, 223
Red-green-blue (RGB), 235–236, 260–261
Reference frameworks, 16
Regression model, 189–198
 for diabetes mellitus, 192–194
 in epidemiology, 194–198
 for heart-related diseases, 190–191
REL. *See* Relation classification (REL)
Relation classification (REL), 86, 95
Relation extraction (RE), 78
 evaluation, 80
 problem, 86
Relative power level (RPL), 164
Relevance analysis, 182–183
ReLU. *See* Rectified linear unit (ReLU)
Renowned convolutional neural network, 106–107
ResNet34 architecture, 5
RGB. *See* Red-green-blue (RGB)
RMSE. *See* Root mean squared error (RMSE)
ROC. *See* Receiver operating characteristics curve (ROC)
Root mean squared error (RMSE), 209–210
RPL. *See* Relative power level (RPL)

S
sALS. *See* sporadic ALS (sALS)

SciBERT, 79, 95
 architecture of, 95
 with BioBERT, 96, 97t
 fine-tuning process, 95
SEE. *See* Shut eye examination (SEE)
Segmentation, heart sound recordings, 219
SEIR. *See* Susceptible-exposed-infected-recover model (SEIR)
Semantic interoperability, 15–18
 benefits of, 21–22
 challenges in, 22–23
 critical security issues of, 19
 healthcare organizations, 19
 in healthcare, use case of, 19–21, 21f
 and limitations, 24–26t
 in patient-centric care, 23
 stakeholders, 22–23
Semantic technology, 15
Semantic validation, 14–15
Sensitivity, classification model testing, 184
Sentence similarity evaluation, 80
Sentences intention label, 95
Serum microrna profile, 137
Serum miRNA expression profile, 134–139
SGB. *See* Stochastic gradient boosting (SGB)
Short-time Fourier transform (STFT), 217
Shut eye examination (SEE), 168–169
Silhouette plot, 64–65
SIR. *See* Susceptible-infectious-recover (SIR) model
Skeletal muscle mrna expression profile, 134–135
Smart grid, 15
Smart mobile phones, 122
Social system groups, 122
Softmax activation function, 111–113, 153
Specificity, classification model testing, 185

Specific natural language processing task, 80
domain-specific tasks, 76, 80
sporadic ALS (sALS), 134
SQL lite, 48
SSI. *See* Stroke severity index (SSI)
State-of-the-art biomedical entity normalization, 80–81
State-of-the-art experimental result, 73, 97–98
State-of-the-art model result, 79
Static node sensor, 124–125
Stethoscopes, 216
STFT. *See* Short-time Fourier transform (STFT)
Stochastic gradient algorithm, 124–125
Stochastic gradient boosting (SGB), 188
Stochastic models, 195–196
Stroke severity index (SSI), 188
Structural interoperability, 17–18
Substantia nigra mRNA expression profile, 134, 140–142
Supervised learning, 180
  algorithm, 147–151
Support vector machines (SVMs), 147–149, 180–181, 187–190
  driver drowsiness detection, 171
  heart sound recordings, 222, 225–226, 227–228t
Survey methodology, 35–39
Susceptible-exposed-infected-recover model (SEIR), 197–198, 198f
Susceptible-infectious-recover (SIR) model, 196–197, 197f
SVM. *See* Support vector machines (SVMs)
Syntactic interoperability, 15–16, 19
Syntactic validation, 14–15

T

Task dataset evaluation metrics, 94, 94t
TCGA research network, 109–110
TCIA. *See* The Cancer Imaging Archive (TCIA)
Technical integration (technical interoperability), 16–17
Text classification task, 96
TFS-Shah method, 3
Three-way complete neutrosophic graph, 240
Three-way complex fuzzy concepts
  algorithm for, 246–247t
  decomposition of, 248–251, 250t, 259–260, 259–260t
  d-equal, 247–248, 249t, 257–259, 258t
  generating, method for, 245–247
Three-way complex fuzzy matrix, 248–249, 250t
Three-way complex neutrosophic concept lattice, 251–257, 255–256f
Three-way complex neutrosophic set, 238–244
  d-equal, 257–259, 258t
Three-way formal fuzzy concepts, 244
Three-way fuzzy attributes, 234f, 239, 244–245, 251
Three-way fuzzy concept lattice, 243–244, 251, 261–263
Three-way fuzzy context, 239, 248–249
Tissue gene expression profile, 131–132
Tissuestwo-dimensional convolutional neural network, 108–109, 108–109t
Training algorithm, 7b
Training data set, 179–180
Training image data, 106–107
Transplantation, 59, 70
True negative, 184

Two-dimensional convolutional neural network, 106
Type 2 diabetes, 58

U

UI. *See* User interface (UI)
UMass memorial medical center, 81–84
U-Net model
  clavicle bone segmentation accuracy, computation of, 210
  computed tomography (CT) segmentation, 205–206, 209f
  construction, 207, 208–209f
  data, 206
  dice similarity coefficient (DSC), 210, 210t, 212f
  network architecture, 206–207
  training and optimization, 208–210
Unlinkable mention prediction module, 80–81
Unsupervised feature extraction, 132
Unsupervised learning, 180
Unsupervised pretraining manner, 79
User interface (UI), characteristics of, 45–47

V

Various biomedical text mining problem, 78
Vascular age, 62–67, 66t
Video recording, 47
Visual odometry techniques, 2–3
VKA. *See* Vocabulary knowledge acquirement (VKA)
Vocabulary knowledge acquirement (VKA), 43–44, 52–54, 54f

W

Water, pH measurement complex fuzzy set, 238–239

Water, pH measurement (*Continued*)
  complex neutrosophic graph, 242–243, 242–243t, 243f
  falsity complex membership value, 253t
  indeterminacy complex membership value, 252t
  three-way complete neutrosophic graph, 240–241, 241t, 242f
three-way complex neutrosophic set, 240
three-way fuzzy context, 239–240
true complex membership value, 252t
Wavelet transform, 217
Wearable sensors, 147
Web technology, 15
Weight matrix, 113
Weight parameter vector, 112
WHO. *See* World Health Organization (WHO)
Wireless capsule endoscopy, 1–2
Words english wikipedia, 75–76
World Health Organization (WHO), 163, 215–216

X
XML, 48

Y
Yawning, evaluation of, 170